Masks, Crutches, and Daggers

The Science of our Addictive,
Self-Delusional
Homo economicus Brain

Ray Armat

Masks, Crutches, and Daggers
The Science of our Addictive, Self-Delusional *Homo economicus* Brain

First published in the United States of America by Applied Analytic Research Inc.

ISBN 9798985969405 (paperback) | ISBN 9798985969412 (ebook)

Library of Congress Control Number: 2022935043
Printed in the United States of America
First Edition 2022

Cover Design by CookieDoh Studios LLC

The information provided is for educational purposes only, and is not a substitute for professional advice. Although the author and publisher have made every effort to ensure that the information in this book is correct, they do not assume, and hereby disclaim, any liability to any party for errors or omissions.

About the Author and the Book

As humans, why do we walk such a fine line between depression and addiction, paranoia and recklessness, fatigue and burnout? Why are humans so prone to self-delusion and self-deception? Why do we have so many conflicting stories and versions of the truth? Why is it such a challenge to give up bad habits?

A graduate of University of Michigan and Case Western Reserve University, Ray Armat, Ph.D., is a materials scientist, chemical engineer, former NASA grantee and educator with years of interdisciplinary corporate research, business development and academic experience. He specialized in balancing industrial product formulas but once he realized his own body and mind were imbalanced, he tempered his ambitious career goals to learn more about the human brain and behavior. As a scientist philosopher, storyteller and educator, he blends real life stories and scientific first principles to find unifying explanations to disease, disorder and discord.

In this book, we learn, in simple language, about the neurochemical soups that make our brains prone to metabolic imbalance, self-delusion, addiction, broken feedback loops and unfettered quantitative growth targets that construct a world of winners, losers and barely anyone in between. Visit RayArmat.com for updates on his next book *Homo economicus: A Guide to Understanding the Present and Future of Our Denatured Species.*

Reviews

"Independence of thought tuned with critical reason and a beating human heart produces the ideas and opinions that change the world for the better. Ray uses his book to introduce us to the world as he sees it filtered through his years of experience, research and accomplishments. What he sees, he shares. This is freedom of thought par excellence. Without it the future becomes the regime of central authoritarians and violence instead of a regime of consent and sharing. Thanks Ray for your contribution to the ongoing conversation. I applaud your effort, honesty and unique and valuable insights. Carpe diem!"

- **TED MCDONALD, aka Barefoot Ted, spiritual naturalist, entrepreneur, 21K ultra-marathoner, featured in C. McDougall's best-seller *Born to Run***

A recent trend in the sustainability literature is assessing the human mind, in particular its shortcomings for dealing with big-picture, long-term challenges. Masks, Crutches, and Daggers takes such assessment to new depths; an impressive display of insightful, transdisciplinary investigation.

- **BRIAN CZECH, author of *Shoveling Fuel for a Runaway Train*, and founder of *Center for the Advancement of the Steady State Economy* (CASSE)**

Dedicated

To independent bloggers, podcasters, journalists, comedians, writers, readers, artists, educators and critical thinkers like you, who have the wisdom to question, and the guts to defy herds, mobs and gatekeepers.

And,

To my family who taught me patience and contentment with simple joys in life: Nature, music (banjo, violin), books and good food!

Warning:

I have been told this book may evoke a strong response in some readers. I have tried to be scientifically objective and respectful of all views. Yet in staying true to the message of the book, I cannot sugar coat my observations. Nobel Prize laureate Albert Szent-Györgyi once said "The real scientist is ready to bear privation and, if need be, starvation rather than let anyone dictate to him which direction his work must take." Today self-censorship is the main form of censorship. If you are looking for a book that confirms your biases, this book may not be for you!

Table of Contents

My Epiphany, in Palm Springs at Frank Sinatra's House!

Imbalance and fatigue are two sides of the same coin and so prevalent in human lives that we have become blind to them. Today, humans suffer mostly from abuse not by others but by their own brain. The world's largest battles are fought inside human minds. We are the only species metamorphosed to have the brain in the driver seat of our body-mind axis. Yet most of us are more knowledgeable and updated about our phones, apps, games, pets, homes, cars and jobs than about our own brain and body, and how they work (or get burned out) together.

I keep coming across posts like this in social media *"It is one of those days that even my second pot of coffee is not helping!"* which now reflects a widespread sentiment. As coffee beans are becoming, as a natural stimulant, one of the most traded commodities in the world, sound sleep is becoming a rare commodity in human lives. The markets for artificial stimulants, alcohol, antidepressant, anti-anxiety, and painkiller drugs have also skyrocketed[i]. The majority of people I know seem fatigued, in pain or feeling stuck in a rut. The Covid-19 pandemic did not cause our fragility. It just uncovered it. And you know the rut is official when, in the age of social media and constant contact, governments in countries like the UK and Japan now have Ministries of Loneliness. Even self-help gurus are not spared. Recently, one of the world's top anti-addiction self-help psychologists almost died after getting hooked on an anti-anxiety drug, and a young billionaire author of a bestselling *self-help* book about happiness became suicidal and reckless, and died inside a mysterious fire[ii].

Unlike other species, human evolution is increasingly leading to the formation of winners, losers and barely anyone in between, financially, that is. Many of the world's largest cities like Los Angeles and New York are now experiencing sharp concurrent increases in both the number of homeless camps and the number of expensive vacant or seasonal homes.

In search of solutions, people who have lost their faith in religions now worship at the altars of business and science, in the hope of finding the fountains of happiness. The term *Homo economicus,* loosely describes the new (species of) human brains that can only be motivated and rewarded by unfettered technological and economic expansion.

---

[i] U.S. distillers' 2020 liquor sales broke a new record and were 8% higher than 2019 despite bar, club and restaurant closures during the pandemic. Whiskey & vodka were by far the leaders in market share increase.
[ii] Details shared in chapter 1

As a scientist and businessman, once trained by alpha *Homo economicus,* I can attest that science and business do not have all the solutions. My own life experience taught me that our problems are rooted in the unique evolution of our brains, which are easily prone to imbalance and self-delusion. I begin with my own story so that you understand why I am writing this book. I am hoping that you find useful insight and inspiration in it.

I was the first person in my extended middle-class family to finish college and graduate school on full merit scholarships. After finishing high school, I spent 13 years in several universities – studying materials and polymer science, chemical engineering, business administration, finance and marketing – and 16 years in the corporate world or working on government (NASA) grants learning how to design chemical and physical systems. I have also taught college courses in science and business. I was one of those behind-the-scenes scientists/engineers who formulate materials used in your shampoo and toothpaste, bread bags, coffee packages, children toys, as well as medical, automotive and aerospace applications. Like many other subject matter experts, I knew a lot about my areas (materials and business) but almost nothing about the broken feedback loops in my brain and body that kept me imbalanced. I worked hard, made money and spent it to buy happiness. I thought I was running a tight ship. But my ship was tipping, actually sinking, all the while I was adding more fuel (money) and running full throttle to pass other tipping ships towards who knows where. I was productive but numb, a state neuroscientists and psychologists call *perseverative.* Busy repeating certain habits but ignoring feedback loops. I was sliding down an unhealthy spiral, like that proverbial frog in a boiling pot, Until ….

****

My pressing of the pause button did not happen because I read Frost's *The Road Not Taken* or Thoreau's *Walden.* My journey started when I was invited, some 20 years ago, as a guest by a corporate client to a corporate rest and relaxation (R&R) retreat at Twin Palms, once Frank Sinatra's lavish exclusive hospitality resort in Palm Springs, California. The charming complex was then owned privately by one of my corporate clients[iii]. I was incidentally housed in the famed cozy bungalow where John F Kennedy and Marilyn Monroe allegedly spent the night. I also dined at the hangout for the Rat pack (Frank Sinatra, Dean Martin, Sammy Davis Jr, Joey Bishop, and Peter Lawford). I could not help but feel hubris and pride. My hard work, competitive drive and education had finally paid off. I was now rubbing elbows and playing golf with top corporate executives, designing and marketing products for major industries, and managing to increase my income by about 30% a year, not an easy feat in the overly competitive setting of corporate America. I felt like singing aloud Sinatra's "*I did it my way!*"

[iii] Now owned by Natural Retreats and available at steep nightly fees to corporations and wealthy individuals.

Yet, deep inside, I knew, like Sinatra purportedly knew about his life, that my polished successful image was *a mask* and my busy life and profitable career *a crutch* to distract myself from an otherwise anxious life at home: a marriage falling apart, a body mass index of 28, headaches, allergies, short temper, you name it. And I was not alone. Most of my colleagues – corporate managers and directors – also lived agonizing traumatized lives: broken marriages, estranged, troubled or ill children or spouses, alcohol and pharmaceutical dependence, diseases, hospital bills, loneliness, etc. If the road to financial success was a highway, it would have been shut down long ago for its casualties and risks.

During the tour of Sinatra's master bedroom, which was equipped with bulletproof windows, I was surprised to explore a hidden emergency underground escape route, which he could use if and when his enemies came after him. This was a sure sign of an anxious, unsettled, sleepless life for a financial and sexual icon that was oozing confidence and worshipped by the world. The man who *did it his way* and symbolized success to the world, up close, inside the lonesome privacy of his home-sweet home, lived a life of fear and anxiety. By the time of his death, Sinatra had divorced several times, suffered anxiety and various illnesses for the latter half of his life, and once even attempted suicide by putting his head in the oven and turning on the gas.

Sinatra's relentless drive to fame and fortune, his impeccable-looking estate and perfectionist attitude were *masks* to his tumultuous life of imbalance and insecurity. He was extremely self-conscious about his looks and used a lot of makeup. He hated to be photographed on his left side. The physical insecurities didn't end there. Sinatra also wore elevator shoes to boost his 5'7" stature up a few inches. He was known to have a violent and unrestrained temper. One of the original bathroom sinks in Twin Palms had a crack in the basin from a champagne bottle that Sinatra hurled at Ava Gardner (his second of four wives).

Imbalance was very costly for Sinatra to maintain. It wasn't the Alzheimer's that killed him, it was the bourbon, according to Mitchell Fink[iv]: "*Sinatra was drinking a bottle of Jack Daniels a day until he was 80 … When his doctors told him that he needed to stop drinking, he did what a lot of alcoholics do in similar circumstances: He switched to wine.*"

And it wasn't only Sinatra who struggled with various addictions. Not far from him lived one of the most famous and powerful couples in history: Lucille Ball and Desi Arnaz. After their divorce, Arnaz confessed to his life-long struggle with moderation in business and life (drinking, gambling and skirt chasing). There are countless other stories about celebrities plagued with traumatic lives, damaged relationships, health issues and addictions lurking behind their toothpaste smiles and masks.

****

After returning from that trip, I wondered why humans struggle with moderation and what makes us so prone to vacillate between extremes of

iv Johnson, Richard. "Frank Sinatra Drank a Bottle of Jack Daily," *Page Six*, December 10, 2014.

depression and addiction. I felt parts of my driven and anxious life mirrored those of Sinatra's. What afflicted Sinatra does not ail only the rich and famous. Bad habits, addictions and depression pervade the lives of many of my working class friends and family.

As a society, we think working hard could make us enough money to solve our problems. In social matters, we point fingers at politicians for not spending enough money on this crisis or that. But spending money alone cannot solve our problems. If it did, rich folks would never be depressed, unhealthy or anguished. Even if money did help, governments are now running out of it due to the sheer number and depth of crises. Imagine our planet as an out-of-balance, heavily-loaded, energy-inefficient sinking ocean liner, ridden by disease and disorder, and we, as passengers, also energy-inefficient and imbalanced, are infighting over burning more fuel (money) and over the ship's destination. The ship is sinking because we have forgotten that the common denominator of disease, discord and disorder is *imbalance* and *metabolic inefficiency*. And because imbalance leads to injustice and trauma, in the long term it becomes costly, very costly, to maintain.

I would like to preemptively ask you to indulge me for my copious use of words like balance, feedback loop and metabolic efficiency throughout this book. As explained in chapters 2 and 3, these principles are the most fundamental laws of nature, from subatomic all the way to interstellar scale, upon which all branches of science and our civilization have evolved. Yet we have departed from them in our own evolution and have long forgotten their significance in our daily lives.

How are we then evolving and adapting if not bound, like other species, by natural balances and feedback loops? I make a case in this book that we have adapted to human-made selection pressures by using three adaptive tools: Camouflages and personas (masks), gadgets to patch our trauma (crutches) and gimmicks and cunning mind games (daggers).

The more we master these tools, the higher we can move up in the increasingly steep sociopolitical power hierarchies we have created. Between 1991 and 2001, I increased my income at an average pace of 30% a year. But I could not hold a candle to folks I used to admire, the mercenary executives and businessman-scientists whose bonuses alone were more than 50% of their base salary. I once met the powerful CEO of a Fortune 50 company and as I was shaking his hand, looking into his eyes and peering into his soul, I could not help but think of the movie character played by Al Pacino (as the devil incarnate) in *The Devil's Advocate*. Everything about this guy was seamlessly planned and calculative. For years, I admired and was mentored by people like him who control the flow of power and money in our world's current dominance hierarchy.

But after the visit to Sinatra's house, I had more empathy for the likes of him and also knew I did not want to end up like him. I needed to balance my life. I was not sure where to start.

Like most people, I sought insight from experts, from doctors on my allergies, weight gain and recurring kidney stones, from counsellors and psychologists on my anxiety and broken relationships, and from self-help authors and influencers on the root causes of my issues. I found some good advice but mostly about the symptoms and not the roots, and the advice often came from doctors who seemed unhealthy, marriage counselors who were single or divorced, and self-help gurus who themselves vacillated between depression and addiction. My fellow scientists and businessmen could be no guides either as they were more or less on the same path I was. I needed real answers and unifying theories so I thought why not start from the fundamental laws of chemistry and physics that govern everything including human brains and behaviors?

In my PhD. program at the University of Michigan they taught us how to use a systemic framework to study mechanisms underlying our observations. So now I was faced with a difficult task of modeling humans as a system to explain the broad biochemical and psychological mechanisms that have left us intoxicated with economic and technological advances but also mired in anxiety, depression, disease, and handicapped by masks, crutches and daggers.

A few months after my dreamy stay at Sinatra's house, I was singing Johnny Cash's *"What have I become, My sweetest friend?"*

****

In chemical and engineering terms, a well-functioning system is said to be in a *steady state*, which means the flow of materials and energy in the system, however complex, is such that dynamic equilibria are achieved across all elements of the system but also inside each element such as reactors, pipes and pumps. Without reaching this state of balance, most of the industrial products we use in our daily lives would be extremely costly and even impossible to manufacture. Without a steady state, your coffee beans or grounds would have a different coarseness (particle size distribution) in each new batch you purchased from the store, hence variable brewing, percolation time and taste[v]. Your home's electrical power would surge and damage your appliances. Without steady state processes, your pet food would look and taste different each time, and you would get a different mileage per gallon for each tank of gas. And we would have explosions, power surges, shut downs, leaks and disruptions in all kinds of chemical and distribution processes. Life would never be the same if engineers and scientists did not manage to create steady-state self-balancing systems and processes. Actually, our solar system, and life as we know it, would collapse absent a steady state, i.e., long-term self-adjusted dynamic balances. Yet, humans have lost steady states at all systemic levels: Superorganism (global), colony (social), organism (individual) and organ (body parts) levels.

---

[v] I worked for about a year on a research project to improve the shelf-stable packaging of fresh-roasted coffee, visited several coffee roasters and smelled like coffee that whole year!

There are numerous insightful philosophical, religious and sociopolitical books that address humanity's traumatic burdens. In this book, however, I primarily use evolutionary biology, neuroscience, biochemistry and physics to describe human bodies and minds. And because this is a book about balance, and because I am not political, I will avoid politics and rely on objective science. I will also be guided by my life experience as well as my conscience, not by an urge to sell any products or romanticize any ideologies or doctrines. This is a book for people who are driven both by career goals and finding balance and peace. Here are the Epigraphs:

In Chapter 1 "*$25,000 for a $2.95 Hot Dog! Why Natural Balance Matters*," I discuss imbalance as the common denominator of disease, discord and disorder but also economic progress. Imbalance is costly because it means constant strain, socioeconomically and biologically. I provide real stories to reflect on the grotesque, complex, unwieldy and cost-prohibitive system we have built and how imbalanced we have become as a species. I contrast humans' metabolic imbalance/inefficiency with the 720,000 miles/gallon of transportation energy efficiency in some other species.

In Chapter 2 "*From Diapers to the Butterfly Effect: A Simple Primer on Thermodynamics and the Science of How Organisms and Systems Work*" I will offer a primer on the most fundamental laws of nature (and science) that will help us understand, in simple terms, how nature self-balances itself over time and space. Feedback is exchanged at all levels and its flow cannot be easily stopped in nature. I will share examples of the cost humans incur each time we try to bypass or fight nature's unfiltered balancing feedback loops such as molecular level colloidal forces.

In Chapter 3 "*Lichens, Cannibalistic Mating, SCOBY, Methuselah and Fibonacci; What Everyone Should Learn from Natural Evolution*" I describe how natural evolution is an intelligent process in that it ensures collective resilience and continuity of life from a gene-centered view, across species, time and space, i.e., ecosystems. This continuity is the result of a process which balances local chaos with long-range order (as in fractals), symbiotic evolution with individualistic adaptation, and within-group (competitive) selection with between-group (collaborative) selection. I describe the mathematical algorithms and wisdom behind natural patterns such as arborization, which is shared by trees, our brain neurons and pulmonary bronchi. We will also learn that natural evolution rewards moderation, biodiversity and metabolic efficiency and punishes excess and monopoly (such as monocultures).

In Chapter 4 "*The Chemical Soup that Controls Human Brains; What Everyone Should Know about the World's Largest Battleground*," I simplify and summarize the latest neuroscience behind human motivations, habits and behaviors. Human brains originally evolved to budget the energy needed for the so-called 4 F's of survival: Food, Fight, Flight and F..k (reproductive sex). However, when human brains developed an ability to imagine and conceptualize, they bifurcated their host's (human's) evolutionary path from that of other species. This first became a blessing but then a curse. When the brain moved the body out of the driver seat, metabolic energy efficiency was no longer an adaptive or balancing trait. As a result, as humans we constantly create ecosystems that easily condition our own brains into self-reinforcing (broken) feedback loops leading to self-delusions, addictions and behavioral ruts such as endless reward seeking, irrational fear (paranoia), recklessness, greed and narcissistic rivalry. I also show how metabolic inefficiency and imbalance (allostasis) is physiologically the root of most evil (disease) in our body.

In Chapter 5 *"A Brief History of Humans and Their Brains: The Fruit that Ushered in the Self-Delusional Homo economicus Brain"* I study in parallel the timeline of our socioeconomic evolution (history) with our brain's evolution (neuroscience) to illustrate the gradual transition of human brains from a metabolic to an economic mode that ignores natural feedback. I will share a brief history of humankind and interesting factoids such as how one fruit changed the path of human evolution and helped *Homo economicus* subjugate *Homo sapiens*. Because our modern concept-driven brain often dwells outside current physical space and time, we are motivated or fear-conditioned by perceived (non-metabolic) stimuli and are never too far from self-delusions, addictions and irrational prejudices. I share shocking results of interesting psychological studies about our self-deceptive brains.

In Chapter 6 *"Masks, Crutches and Daggers,"* I explain how calculative economically-wired brains are helping humans metamorphose into *Homo economicus*, a metabolically inefficient species handicapped in processing self-correcting natural feedback loops. As a result, we now rely on an adaptive toolkit consisting of masks, crutches and daggers. These may be beneficial in the short-term but further alienate us from nature's balancing and self-correcting feedback. I share how reliance on masks and crutches did not evolutionarily bode well for crafty intelligent species like cephalopods.

In Chapter 7 *"Our Schaub-Lorenz TV Needed a Kick! Good vs. Bad Science: The Lobotomy Orgy, The Omnipotence Paradox,"* I share thought experiments to show how our numbers-conditioned brains create scalable economies and reductionist sciences which thrive on problems (imbalance) and not solutions (balance). This is in contrast to adaptive design principles used in nature based on system wide (wholesome) metabolic efficiency and balance. I will discuss how our modern scalable economic approach has impacted our health and modern sciences.

In Chapter 8 *"Our World is Our Mirror: Paths to Balance, Peace and Health,"* I explain how the silver lining to our brain's vulnerabilities is its neuroplasticity and our ability to reverse-condition it back onto a path closer to its natural evolutionary roots. In the final chapter of the book I combine science, philosophy, stories of mystics, and practical personal anecdotes in search for clues about re-conditioning our brains. The content is more or less free flowing and reflects the open-ended nature of the choices ahead of us, as a species which has lost its evolutionary trail in nature.

Masks, Crutches, and Daggers

****

I believe independent-minded souls brave enough to question the wisdom of unfettered quantitative growth will be harbingers of balance for our species and our ecosystem. You, the independent-minded reader, are honoring me by trusting your time to me.

Chapter 1: $25,000 for a $2.95 Hot Dog! Why Natural Balance Matters

Epigraph: What is the common denominator of disease, discord, disorder? Imbalance! As a species, our ability to ignore or bypass natural balancing mechanisms and feedback loops has helped us achieve economic growth in terms of GDP (gross domestic products). But in the long term, imbalance yields to a complex and unwieldy system, which is costly to the poor and rich alike because it means constant strain, socioeconomically and biologically. I provide real stories to reflect on the grotesque and cost-prohibitive system we have built and how imbalanced we have become as a species. I contrast humans' metabolic imbalance/inefficiency with the 720,000 miles/gallon of transportation energy efficiency in some other species.

For $350,000 a Month Become Tom Hanks' Neighbor

Recently the Food and Drug Administration (FDA) approved a drug designed to delay the onset of Alzheimer's disease[i]. The drug costs $56000 a year per patient to delay the symptoms of the disorder. Around the time the drug was approved, news headlines showed some 700 cars waiting in a 2.5-mile long food bank in Miami, Florida to pick up the $20 food packages they badly needed. Also, at the same time, on the west coast, a house perched atop a promontory overlooking the Santa Monica Mountains and the ocean, sitting next door to a mansion owned by Tom Hanks and Rita Wilson, was rented for $350,000 a month as a retreat for rich folks escaping virus-plagued cities and food lines.

I am a follower of news, not the politics of it but the *math* and *numbers* behind it. So after a simple calculation, I realized the rental cost for that mountain-top villa was about the same as the cost of feeding the 700 families in that food bank for a month or the cost for a wealthy person to delay his or her Alzheimer's for about six years. These are all ways our species is allocating its resources and budgets.

I enjoy many of Tom Hanks' movies. I also hate to see people suffer from Alzheimer's or wait in foodbank lines. This book is not about the popular act of casting stones at the rich either. I am sharing these stories and other examples in

---

[i] A disorder which may be rooted in modern day's prolonged stress-induced hormonal imbalances

this chapter in order to make it hard for anyone, even those who admire technology, modern life and advanced economies, to deny the level of gross imbalance and disparity we have built into our systems. The stories of this chapter illustrate the imbalances built into the humankind's new (artificial) evolutionary path, which is bizarre when measured by yardsticks used in nature for the evolution of other species.

What Do Bowerbirds and Bill Gates Have in Common?

Let me share an example to demonstrate how different we have become from other species. We know very few species naturally build nests or dens that are much larger than their physical size or footprint. This is to remain efficient by conserving materials (mass) and energy and remaining free to mobilize, if needed. Yet there are rare examples of excess and exuberance in nature. Take the case of male bowerbirds in the forests of New Guinea and Australia which are known to spend days building elaborate colorfully decorated courtship nests (called bowers) in response to sexual selection pressures. The male bowerbirds basically show off their bower building skills and stamina as an evolutionary signal of fitness to attract females for mating.

But even bowerbirds do not come close to humans in exuberance and excess. Typical bowers are about 5-10 times the size of the bird, as compared to human abodes and mansions that can fit tens or hundreds of humans. Bill Gates' main house, one of his many, spans over 66000 square feet (6600 square meter), and contains 37 rooms, 27 bathrooms, a 23-car garage and a reception hall that accommodates 200 guests. Reportedly, a special grade of sand, imported from the Caribbean islands some 5000 kilometers away, decorates their private beach. The building of the house used up about half a million tree logs[1], which were once habitats of numerous animals and species.

It took the builders seven years to build Gates' house but the real cost, measured in terms of the productivity (income) of an average *Homo sapiens* (human) is more than a thousand years. Compare that to the most elaborate nests in nature. It takes an average bowerbird no more than a couple weeks of labor (with the purpose of lovemaking) to build a fancy bower from local materials. The Santa Monica mansion in Tom Hanks' neighborhood mentioned earlier costs, in human's productivity (income) rate, some 1000 years to pay for. Even small human abodes most of us own cost an average of 3-5 years in productive labor output. With such high costs spent on our residences, no wonder humans are so immobile and attached to their fixed assets.

Nature could not even start to make sense of the human economic calculus and trade-offs[ii] because in nature, species and objects only exchange materials, momentum and energy so they need to be extremely efficient with all[iii].

[ii] That is old economics. The Modern Monetary Theory (MMT) based on unlimited debt issued by governments is not even understood by most humans!

[iii] There are rare examples of trade in nature such as food-for-sex trade among some Chimpanzees

But what is so unique about humans that allows us to use up so much material and energy so inefficiently and without being penalized by nature? What allowed us to separate our evolutionary path from the natural Darwinian path and selection pressures impacting other species?

As will be detailed in chapter 4, thanks to unique cerebral abilities of conceptualization and abstraction, humans can set their own boundaries and limits and construct (unnatural) worlds by bypassing normal feedback loops that impose boundaries and balance in nature. But by doing so, humans have become metabolically wasteful and inefficient. Let me explain how.

Natural vs. Human Evolution: 720,000 vs. 30 Miles per Gallon

Our distant ancestors, *Homo erectus* and early *Homo sapiens*, lived in a naturally balanced world. If you are not familiar with the theories of evolution, I will provide a summary in chapter 3.

The theory of natural evolution is based on a constant competition between and among species to survive and flourish in the face of limited natural resources and bottlenecks (selection pressures) in their ecosystem. Those genetic variants within a species that can adapt better to selection pressures of their ecosystem and win the arms wrestling over resources with other variants or species, can secure enough food, sex (reproduction) and shelter to survive and pass on their genetic code, i.e., be selected for posterity. That means a rough and wild world in which individual species are selected for physical traits like speed and strength but everyone's level of consumption, procreation and aggression is kept in check by the necessity to remain resilient and efficient. The three fundamental laws of nature, namely the conservation of mass, energy and momentum (Chapter 2) ensure species become efficient in mass, energy and momentum (force) balance and allocation of resources within the ecosystem. Balance becomes an adaptive trait because excessive (greedy) members of any species would either run out of food or become overweight, slow, unfit and soon food for others.

Mathematically, this is why natural evolution and competition over limited resources mean polygenic traits in populations often follow a classic Gaussian bell curve distribution. There are very few extreme traits in nature. Even with alpha-ranked animals high on the dominance hierarchy, the amount of energy or food they consume is often only slightly higher than subordinate animals. For example, in schools of trout, the metabolic cost of maintaining an alpha position in terms of swimming or aggressive behavior is no more than 25% of the daily caloric needs[2]. Furthermore, unlike in human hierarchies, alpha animals in nature often pay a high metabolic stress cost (cortisol/testosterone) that will lead to immunosuppression and shorter lives. Imagine if Bill Gates had to metabolically spend most of the calories from his daily food intake on fighting competitors, building his house or protecting it. He would find himself a hungry and short-lived man.

Nature often discourages, rather punishes, extreme vanity. Darwin's classic example of balance in natural adaptation is the case of the peacock's tail. Male peacocks with larger, more colorful tails will attract more females, so by mating and reproducing more, their genes will be sexually selected. However, the natural selection process sets limits to this trait and prevents exuberance and unfettered excess. Overly pompous peacocks with too spectacular of a tail will attract more predators than mates and end up as food. So from a metabolic standpoint, a super tail is maladaptive because it uses too much energy for fighting or avoiding other peacocks and predators. Sexual selection in nature often does not prefer Casanovas. Extreme opulence would not be adaptive in nature, which achieves biodiversity and ecosystem-wide metabolic efficiency by rewarding balance and harmonized (symbiotic) cohabitation.

Species in a Darwinian world are often mean, lean and tough. Adaptation is governed by physical traits and selection pressures in local habitats. For example, some sea creatures, like the giant isopod, can go for years without eating. This is an adaptive trait in their deep ocean habitat where food is scarce. Others, like Antarctic or Patagonian toothfish evolved a slow heartbeat (of 10 beats per minute) and antifreeze proteins in their blood and tissues to prevent freezing in subzero temperatures.

Now let's have a critical look at the new evolutionary path humans have chosen in which balance, whether self– or naturally imposed, is no longer an adaptive trait because we have developed, through collaboration and competition, superior intelligence, tools and rules that allow us to ignore, block or divert natural feedback loops. By bypassing the checks and balances of natural selection, *Homo sapiens* practically have access to boundless speed, quantity, reach, convenience, opulence and comfort. We live in the proverbial "Sky's the Limit" type of world. We could now medicinally or genetically customize a peacock's tails and his sexual drive. Our own opulence and extremes are not bound by nature's limits. One of my favorite poems is by the mystic poet philosopher, Rumi, who urges kings and queens (the rich and famous) to learn from the destiny of peacocks[3]: "*Many a peacock pays a price for his opulent plumage, Many a king for his opulent crown.*[iv]"

But Rumi's advice was from 800 years ago. Today's humans can not only beat peacocks in opulence, but in flying. Using engineered gadgets, we can now fly more than any peacock or bird can. I used to have corporate colleagues who were proud of membership in airlines' million-mile frequent flyer clubs. I was envious of the perks they received like priority seating and club reception that airlines dangled like a carrot to us corporate travelers. But none of my high flying colleagues could hold a candle to Tom Stuker, a sales training specialist from New Jersey who was just about to celebrate flying 22 million miles last

---

[iv] In fact, ancient Egyptian deities such as Osiris wore crowns embellished with curly red ostrich feathers that symbolized opulence and virulence.

March before the pandemic hit. No birds could fly that much in a lifetime because natural feedback loops ensure and require metabolic energy efficiency.

In nature, radar studies have been able to track the remarkable 3000 km (1800 miles) transoceanic migratory flight of some songbird species from Nova Scotia and Cape Cod heading southeast on a course toward Africa. The flights sometimes last about four days and nights and impose a huge metabolic cost on the small birds. Birds like blackpoll warblers cannot land on water without drowning so they lose (burn as fuel) about half of their body weight – usually around 10 grams preflight – in their non-stop flight to their destination. Imagine my colleague or other frequent flyers losing half of their weight each time they flew.

As humans, we exploit natural resources to our advantage so we do not have to be individually energy-efficient. In fact, we are extremely energy-inefficient when we use fossil fuel to fly. Studies show[4] that flights by blackpoll warbler represent a gasoline-equivalent fuel efficiency of 720,000 miles per gallon[v]. Compare that to human flights and the airline industry's average fuel efficiency per revenue passenger of 58 miles per gallon. Think about these numbers the next time you hear contentious debates over climate change, pollution and environmental protection. Political debates often ignore our metabolic inefficiency as a species which has parted its evolutionary path from others.

We have developed tools that allow us to shunt or ignore limits set by natural feedback loops, like our circadian rhythm, which is disrupted by jet lag each time we fly over several time zones. I know when my frequent-flyer colleagues felt jet lag, they had many choices in their toolkit: Sleep aids, alcoholic spirits, coffee, carbonated soda and energy drinks.

And when things go off-balance, we have developed specializations to patch each other up. If the stress of work and life in Sky's the Limit world gives us high blood pressure, our specialists have just the right prescription to manage our hypertension so we can stay productive. Down the road, if the first medicine boosts our blood sugar (glucose) levels, which is common with many blood pressure medications, we ask our toolmakers for a new solution (anti-diabetics). If these prescriptions over time give us heartburn, there is a heartburn pill we can take too. If pains, kidney, liver or memory issues ensue, we ask our specialists for more fixes, and they oblige.

Man-made tools and rules exploit natural conservation laws (Chapter 2) to our species' advantage, so maintaining an individual-level balance and physical fitness are no longer needed as adaptive evolutionary traits. An injured fox may not survive for long in the wild but in the human world, any of us, regardless of how traumatized (imbalanced) we are, can move up the power hierarchy.

---

[v] Even if we account for weight differences, the warbler's fuel economy is more than 100 miles per gallon in equivalent human terms

The Brain-Damaged Suicidal Girl Who Turned Millionaire

I already shared the story of Frank Sinatra, and how despite all his insecurities and traumatized life, he became an alpha human. What comes next is the real life story of an alpha woman, who despite an extremely traumatized childhood, managed to become famous and rich, very rich, while living a seriously imbalanced life at the same time.

Born to immigrant Russian parents in Salt Lake City, Utah, she suffered from Bell's Palsy– a nervous disorder causing facial paralysis, often induced by viruses, chemicals or drugs – at age three. The family had a rough start in America. Her father made his living as a door-to-door bible salesman, compelling the family to keep their Jewish heritage a secret from their community. She apparently lived in an abusive household and later accused her parents of abusing and molesting her. She reportedly developed a "multiple personality disorder"[5] – now referenced as Dissociative identity disorder, DID, in The Diagnostic and Statistical Manual – 5th Edition. At age 16, she attempted suicide by jumping in front of a car, resulting in a traumatic brain injury (TBI) and a nine month commitment in a psychiatric hospital. While admitted, she gave birth to a child, which she put up for adoption.

In a natural Darwinian ecosystem, she would probably not survive unless she healed her trauma early in life. But humans have the luxury to ignore natural feedback so she learned how to flourish in an economic system that leveraged her imbalances[vi]. She masked her trauma through acting, on stage and offstage, worked hard, used numerous psychiatric drugs (crutches, as defined in Chapter 6) and various ploys (daggers), such as deception and lying, to reach an astonishing level of fame and fortune as a comedian, actor and an iconic influencer for millions of Americans.

As she made more money, she could afford more drugs and specialists to help her with blunting her own body's feedback such as pain, anxiety and depression. She leveraged her imbalances and treacherous fractured personality in a limitless economic system that prized her competitiveness. Her fans cheered her on and turned a blind eye to her trauma. It was a perfect self-reinforcing (unhealthy) feedback loop (natural ones are self-balancing).

Yet over time, her imbalances (trauma) caught up with her. She had fallouts in personal and business life, scandals caused by racial comments, multiple divorces, surgeries, suicidal thoughts and a heart attack. She finally reached a dead-end and started a long overdue search for help to balance her life and heal her trauma. Her name is Roseanne Barr. She once confessed to Oprah Winfrey, *"I was very unhappy .. and I was prescribed numerous psychiatric drugs ... to deal with the*

[vi] As we shall see in chapter 7, a good part of human economic system is now built on deficits and imbalances

fact that I had some mental illness ... I totally lost touch with reality." Throughout this book, I refer to what she calls reality, natural feedback loops and balance.

If you are an ethicist, you may condemn the economic system for exploiting Roseanne's misery. If you are a moralist, you may blame Roseanne for her indiscretions and opportunistic personality. Yet as shown in the book, Rosanne's plight was simply a natural outcome of the new unnatural evolutionary path we have chosen as *Homo sapiens*. Roseanne is not unique in the way our brains work. When she talks about "losing touch with reality" she is describing a neurological condition that we are all prone to, called self-delusion, described in chapter 5.

I hope after reading this book, you agree with me that ethically and morally judging other humans is futile without understanding how we are all prone to self-delusions and screw ups. We are an imbalanced and denatured species.

The examples in this chapter help us realize the huge cost associated with managing imbalances, while they are becoming more grave and unmanageable. There will be more examples in Chapter 7.

Our Prisons: $25,000 for a $2.95 Hot Dog

The roads to prison start years earlier from traumatized childhoods. Yet the resources we allocate to healing childhood trauma is not in balance with the money we spend on prisons.

The tragic story of Joe Ligon is one testimony to the bizarre, costly and ineffective system we have built. Joe was recently released from a Pennsylvania state prison 68 years after he was imprisoned at the age of 15. Eastern State Penitentiary, where he was once imprisoned, is now a museum and Halloween attraction. He abandoned school in the third grade and was illiterate when he was first arrested at age 15. His detention plus the legal court processes cost the human society an estimated $3.5 million (in today's dollar value), a lot more than it would have cost to help his family raise him with a balanced lifestyle and good education.

Prisons across the world are packed and costly to maintain so in many places, despite the public outrage, prisons are handed over to private prison corporations that count prisoners as profitable labor and assets. Fifteen states in the US are now spending $27,000 more per person in prison than they spend per student in school. Some US states such as New York are now spending nearly $70,000 per year on each prisoner. That is more than twice the median income of an average American citizen outside the prison!

To see how imbalanced our priorities are, consider the recent arrest of a young, hungry man in the city of Nashville, Tennessee for snatching a $2.95 package of sausage from a local Walmart store[6]. Interestingly, it costs the state around $25,000 a year (in addition to court costs) to keep him inside the prison if he is convicted. That is about the same amount of money that could feed and house him outside the prison and in dignity.

Another tragic story is of Ricky Jackson who once held the ignominious record for the person who served the longest time in the US penal system before he was proven innocent and released. He was convicted in 1975 for the murder of a white businessman in Cleveland, Ohio, based on the false testimony of a 12 year old boy. He spent 39 long years on death row before he was released in 2014 with the help of lawyers and law students from the Ohio Innocence Project. His detention plus legal fees cost the society an estimated $1.5 million (in today's dollar value). When released from prison, Ricky Jackson said: *"Prison in America is a business, inmates are just commodities. People are being exploited by the thousands, hundreds of thousands."*

As we will see in chapter 7, there are now 36 million lawsuits filed in the United States each year and $650 billion spent on our civil, criminal and military justice systems. Prisons alone in the U.S. now generate about $10 billion a year in revenues. Private prisons alone are now an estimated $5 billion industry. According to Business Review at Berkeley, aside from the federal prison industry, state-run prisons generate millions in profits, making prison labor an industrial complex worth over $1 billion.

This is what an imbalanced (traumatized) society looks like. We spend way more resources on fighting and containing disorder than preventing imbalance. Imbalance is costly and ineffective yet as I will explain in chapter 7, we have evolved not only our bodies but also our economic systems based on deficits and imbalances and not on maintaining a balance.

Our Schools: The Tragic Story of Camden and the Soup Company

The problems facing our schools are systemic as well. The cost of education has been skyrocketing at the same time student performance and educational standards are declining. In some school districts like Avalon, New Jersey, the annual educational spending is now more than $60,000 per student, more than per capita gross domestic product (GDP) of many world economies.

Even the school systems in low income cities are costly. Some 75 miles north of Avalon, in Camden, New Jersey, a low income city across the Delaware River from Philadelphia, Pennsylvania, the school system now spends about $32,000 per student per year, more than the annual household income for many families in the poverty-stricken town. Interestingly, Camden is home to the world renowned Campbell Soup Company with nearly $9 billion in annual sales. That would make the city of Camden richer than many world economies once we include the Campbell corporate revenues. Yet Camden is among the most dangerous and poorest towns in America as judged by statistics of violent crime.

This is what a non-symbiotic imbalanced evolution model produces. For about 0.1% of the company's revenues, every one of the 74000 residents of Camden, many malnourished, could be fed throughout the year, at least with Campbell's nutritious soups.

On the other extreme, we now have wealthy folks buying their children's way into top colleges. In 2019, a college admissions bribery scandal revealed a criminal conspiracy by several wealthy celebrities and school officials to influence undergraduate admissions decisions at several top American universities. For example, actress Lori Loughlin and her fashion designer husband paid $500,000 in bribes to secure spots for their daughters at the University of Southern California as fake athletic recruits for the school's rowing team, even though neither rowed competitively. Our educational system is failing the rich and poor children alike while we spend more per student than most other countries.

Our Drugs: Tioga and the Tragic Addiction of the Anti-Addiction Expert

As humans suffer from imbalances, many of them resort to substances to help with the trauma and pain. The surge in street sales of the super painkiller and deadly drug fentanyl has led to thousands of deaths across the U.S. especially in the Western states. As the demand for these substances skyrocket, legislators try to control and disrupt the supply of drugs by labeling them as controlled and uncontrolled substances. These artificial measures on the supply side have led to an ongoing war on drugs.

The futile and costly war took a new turn on August 14, 2019 when police in Tioga, a poverty-stricken neighborhood in Philadelphia, attempted to arrest a man, accused of illegally buying the anxiety medication alprazolam, a schedule IV controlled substance, definitely not the most dangerous controlled substance[vii]. The man resisted arrest and fired back, injuring several police officers. This led to a 7.5-hour standoff with dozens of police officers occupying and crippling the neighborhood. In my estimate, this one ordeal – similar to many others occurring daily across the US – cost the city and its residents in excess of $1 million once we include costs of the force and equipment deployed during the standoff, treating the injured police officers, covering their salaries, disability and pension, court and trial costs, and the long term prison costs for the man who resisted arrest. Billions are spent each year on this ineffective war on drugs.

Tragically, the shootout was over an anxiety drug which is legally marketed and prescribed by doctors some 25 million times every year. It belongs to the notorious Benzodiazepine family of drugs to treat panic disorders, with a global market value of $3.5 Billion in 2019 according to Verified Market Research. It is the drug family which even harmed the renowned psychologist Jordan Peterson, resulting in his disability, induced coma and placement under suicide watch.

[vii] Schedule I controlled substances include heroin but also marijuana and considered high potential for abuse. Schedule V drugs, substances, or chemicals are defined as drugs with lower potential for abuse. Interestingly, many pharmaceutical drugs are placed in safer categories while natural or illegal substances like marijuana are labeled as least safe.

Peterson, himself an expert on this family of drugs and now a staunch critic, received his prescription legally from his doctor. It has cost Peterson's family more than a million dollars to keep him alive using expensive detox and rehab centers around the world.

These two examples, one involving a violent standoff in a poor neighborhood, and one involving a renowned psychologist and influencer, demonstrate how using drugs as our main tool to fix our psychological imbalances is costing millions of dollars in medical or law enforcement expenses, destroying lives, families and neighborhoods, as well as packing courts, prisons, rehab centers and hospitals.

Besides psychiatric drugs, opioid abuse is also a strong biopsychosocial (to borrow the term from psychiatrist George Engel) barometer for the extent of addictive disorder among humans. In 2020, a record high number of 93000 Americans died of opioid overdose, about twice the numbers in 2017. More than 11 million Americans owe their addictions to doctor-prescribed opioids in 2017. The prescription opioids generate some $12 billion in revenues for pharmaceutical companies and 150-200 million prescriptions-visits (worth about another $10 billion) for doctors every year. In addition, in 2017 alone, each case of opioid use disorder and overdose cost[viii] U.S. taxpayers about $455000. And like many other social trends, the problem is scaling up alongside or parallel to the size of the economy.

And the Covid-19 pandemic and strict lockdown policies have only amplified the sales of anti-anxiety drugs and opioids. In chapter 4, I will discuss the neuroscience behind the imbalances that cause anxiety and depression.

Our Medical Bills: Costing Way More than our Food

Currently we spend more than 20% of our GDP in the United States on medical bills[ix]. Considering that only about 70% of the population is in the working age group responsible for the production of all goods and services, our national medical bills cost us about 1.5 days of the adult population's labor productivity per week or about 30% of our working days. Simply put, without our medical bills, we could all work 1.5 days less per week. We are rightfully thankful to technological and medicinal interventions that allow us to maintain our natural imbalances but we do not realize the cost of maintaining our imbalances, disorders and diseases.

Is our management of disease paying off? In the U.S. over the past 50 years, we have added about 10 years (or 15%) to our average lifespan, from 70 to 80 years old. But did all of that come from our disease management system? Probably not! Non-pharmaceutical interventions have played a huge role in our longevity. For example, smoking prevalence has dropped from 45% to 15%

[viii] Including health care, substance use treatment, criminal justice, lost productivity, reduced quality of life, and the value of statistical life lost. Source: https://www.cdc.gov/mmwr/volumes/70/wr/mm7015a1.htm

[ix] A worldwide trend although Europe spends about half of the US in medical bills as a percentage of GDP

between 1970 and 2020. Even if we owe the entire 10 years of our additional life expectancy to medical care, we have traded off 30% of our productive labor with a 15% increase (10 years) in lifespan. More importantly, what is the quality of living for those extra 10 years we have gained?

Some diseases of imbalance are very costly to manage through modern medicine. For example, Alzheimer's disease, a disorder now linked to modern day's prolonged stress-induced imbalances in hormones such as cortisol, and neurotransmitters such as glutamate, is pharmaceutically hard to treat. As previously mentioned, one recently approved drug, touted to delay the onset of the disease (symptoms, that is), costs $56,000 a year to treat each patient. Despite its cost, the drug was approved by the Food and Drug Administration (FDA) on an "accelerated" path ignoring the recommendation by an 11-member advisory committee which voted nearly unanimously against the drug, citing inconclusive evidence that the drug was effective. Three members of the FDA advisory panel resigned over the agency's decision.

The point of these examples is to show that in our new evolutionary path, we have found ways to live with our individual or social imbalances. But these imbalances lead to more imbalances, disease, disorder, and overstressed medical care systems. Costs will only go up from here. I will explain further how "crutches beget crutches" in chapter 6.

The recent pandemic has cost America hundreds of billions of dollars in hospitalization and vaccine programs. This is in addition to the $15000 we normally spend for each citizen's medical (healthcare) bills every year. $15000 is about four times the cost of supplying each citizen, throughout the year, with wholesome organic top quality groceries, plus major vitamin supplements, plus gym memberships, plus two preventative dental hygiene and two preventative primary care doctor visits a year. In other words, if we had a balance-focus system, as in nature, we could have prevented a lot of disease by spending about 25% of our current medical bills in disease prevention. With fewer bills, we could work less, relax more and, as a healthier nation, reduce our medical bills and psychological stressors even further. This is why we pay huge costs when we depart from natural evolution's path of symbiotic and metabolic balance.

Chapter Synopsis and References

We contrasted natural selection in Darwinian evolution with adaptation of humans in their new "Sky's The Limit" evolutionary path, in which physical fitness and balance are no longer adaptive traits. Imbalance leads to disease, disorder and discord. Instead of maintaining balance, humans have developed sophisticated and costly tools and rules to manage these imbalances, often in unnatural ways that keep the imbalances intact or even encourage them. I shared examples from the world of entertainment, prison systems, education systems and medical care systems to show how grave and costly our imbalances have become as a result of our departures from natural evolution.

[1] "9 Facts about Bill Gates' House You Probably Didn't Know about It." *Arch2O.Com*, 22 June 2021, https://www.arch2o.com/tour-inside-bill-gates-house/.

[2] Feldmeth, C. Robert. "Costs of Aggression in Trout and Pupfish." *Behavioral Energetics: The Cost of Survival in Vertebrates*, edited by Wayne P Aspey and Sheldon I Lustick, 7th ed., The Ohio State University Press, Columbus, OH, 1980, pp. 117–138.

[3] "Rumi's Masnavi – Listen to This Reed How It Complains, Telling a Tale of Separations—." *Rumi's Masnavi – Listen to This Reed How It Complains, Telling a Tale of Separations—* http://www.masnavi.net/. My conceptual translation from mystic poem in Persian.

[4] Kreithen, Melvin L. "Orientational Strategies in Birds: A Tribute to W. T. Keeton," *Behavioral Energetics: The Cost of Survival in Vertebrates*, edited by Wayne P. Aspey and Sheldon I. Lustick, 7th ed., The Ohio State University Press, Columbus, OH, 1980, pp. 3-28

[5] Lachmann, Suzanne. "What Is Going on with Roseanne Barr?" *Dr. Suzanne L*, 19 June 2018, http://www.drsuzannel.com/what-is-going-on-with-roseanne-barr/

[6] https://www.scoopnashville.com/2020/08/nashville-man-charged-with-felony-theft-for-taking-2-95-in-sausage-from-walmart/

Chapter 2: From Diapers to the Butterfly Effect: A Simple Primer on Thermodynamics: The Science of How Organisms and Systems Work

Epigraph: Humans explore nature to exploit it, but in that process we disregard and disrupt naturally evolved feedback loops and cycles that balance the universe. To understand what is unnatural and dangerous about our new evolutionary path I will review laws in chemistry, physics and thermodynamics to explain, in simple terms, how equilibrium and balance is achieved in nature through homeostasis, feedback loops and symbiotic metabolic (energy) efficiency. I call these forces nature's intelligence because they allow nature to self-balance itself over time and space. Feedback is exchanged at all levels and not easily stoppable in nature. I will share examples of the cost humans incur each time we try to bypass or fight nature's unfiltered balancing feedback loops such as molecular level colloidal forces.

****

Diseases and disorders are rooted in natural imbalances in our bodies, brains and ecosystems. Yet most humans, even our problem solvers, influencers and decision makers, are not properly educated about fundamental laws that govern nature and humans. I will share what I have learned about natural forces in chapters 2 and 3, and about human brains in chapters 4 and 5. After reviewing these chapters, you should be convinced, like I am, that natural laws are more intelligent than human-made laws because nature constantly self-balances itself and reaches a scale-free state of symbiotic metabolic efficiency. It is not the technology, politics or business but principles of biology, psychology, evolution, physics, chemistry and neuroscience that govern human organisms and therefore, human-specific trauma, injustice and disease. Only by applying these principles at an individual level, we can balance and heal ourselves and our societies. Money may patch up our disorders but could not fundamentally heal us.

So let's review the simple conservation laws that govern all natural phenomena, including our bodies and minds. This is classical Newtonian physics[i]. I believe many of our schools do not properly teach these cool amazing laws and their practical applications in our daily lives and evolution.

---

[i] Subatomic scale quantum physics is outside our scope in this book although I will make a short reference to it later in this chapter.

You can skip many of the details presented in this chapter if they get too technical or too easy for you. We need to all learn about these natural laws that balance our universe, the same ones humans are working hard to override or exploit.

Interparticle and Interatomic Forces of Nature: A State of Balance

Nature, in inanimate (lifeless) form, balances itself through exchange of invisible forces, mass and energy at all levels and scales. Biological life forms, on the other hand, have evolved to actively (intelligently) self-balance or balance each other so they will resist or compensate for all sorts of imbalances.

Let's start from the smallest scale. All atoms and molecules in this universe interact and exchange forces (feedback) through the so-called Van der Waals forces, named after Dutch physicist Johannes Diderik van der Waals. The forces are weak and distance-dependent so the farther the objects are, the smaller the force, but for small particles less than 250 microns in diameter – one hundredth of an inch or about 2-3 times the average diameter of a strand of human hair – the force is cohesive and leads to formation of caked powders.

It looks like nature is telling us, through Van der Waals and similar molecular-level forces[ii], that particles this small are more stable and balanced when not disturbed. Yet, my fellow curious ingenious scientists and engineers regularly spend a lot of time and money to design systems that break apart bulk powders into fine particulate matter. In fact, the whole field of colloidal science, which I studied in my Ph.D. program, specializes in stabilization of fine matter in dispersions.

The pulverizing of fine particles is done to increase the surface area. We know from our high school geometry that reducing the diameter of a sphere by a factor of 10 will reduce its surface area by 100 times but its volume by 1000 times, with a net effect of "increasing" the surface-to-volume ratio by a factor of 10. Increasing the surface-to-volume ratio by breaking down materials is helpful in applications that depend on "surface" transport phenomena such as absorption or adsorption, i.e., surface binding. For example, diapers, feminine hygiene products, baby powders and some cosmetics rely on very high surface area to absorb a lot of liquid per unit weight. Superabsorbent fluffed cellulose pulp fibers used in some diapers will absorb about 12 g of water per gram of dry fiber, whereas new superabsorbent synthetic polymers will absorb up to 1,000 g of water per gram of polymer, useful not only in diapers but also in applications such as hydrogels, contact lenses, and oil spill cleaning materials.

There are many other applications for superfine powders in powdered egg, spices, milk powder, flour (milled and bleached powder, not the coarse whole grain), sugar, modern pantry items, as well as in industrial products such as paints, plastics, rubber, cosmetics and personal care products. Now the problem

---

[ii] There are other molecular-level forces such as hydrogen bonding, capillary, electrostatic, and surface tension forces which are beyond the scope of this book.

with pushing back against nature's feedback (attractive Van der Waals forces) is that even when we are successful in pulverizing powders into unnaturally small sizes, they will naturally tend to flocculate (scientific term for lumping together) over time and resist free flowing.

So as engineers and scientists we spend a lot of money and time to find ways to force fine powders to stabilize (not lump). One trick used is adding to fine powders anti-caking additives such as sodium aluminosilicate (as in powdered salt), sodium dioxide (as in powdered egg solids and beer filtrates), calcium silicate (as in powdered spices) and cellulose (as in grated parmesan cheese, and as wood pulp in diapers). But as with other commercial formulations, the anti-caking additives need to be cost-effective (cheap). In 2016 Walmart was sued for selling cans of "100 percent" grated parmesan cheese that tests showed to actually contain up to 10 percent cellulose (wood pulp) as an anti-caking additive[1].

But the problem is not just the level of the anti-caking additives. It is their impact on our health. Although many anti-caking additives are labeled by the FDA as GRAS (Generally Recognized As Safe) many are suspected of harmful effects especially in high doses or if "denatured." For example, Aluminum (found in many anti-caking additives) is often associated with Alzheimer's and other neurodegenerative diseases in humans.

Baby Powder

Another problem with denatured pulverized materials is that even after their micronization and successful stabilization, these superfine powders such as carbon black (used in plastics, rubber, paint and electronics) and talcum powder

(used in cosmetics and personal care products like Johnson & Johnson's baby powder[2]) when released into nature or absorbed through our body have been linked to cancers such as lung and ovarian carcinoma. In 2018, a St. Louis, Missouri jury ordered Johnson & Johnson, the giant pharmaceutical and personal care products company, to pay $4.7 billion to 22 women and their families because the company's baby powder containing superfine talc-based particles contributed to the women's ovarian cancer and mesothelioma. An appeals court reduced the fine to $2.1 billion. Although the company continued to sell the product, a Reuters' investigation found that Johnson & Johnson knew for decades that its baby powder could be contaminated with asbestos, a carcinogen that sometimes occurs naturally with talc. Asbestos can occur naturally underground near talc. It becomes harmful when it is pulverized into small particles and lodges in human tissues, possibly leading to diseases including cancer and mesothelioma. Johnson & Johnson finally stopped the sale of the product in 2020 after facing more than 16,000 talc-related lawsuits nationwide. Recently, the U.S. Supreme Court refused to hear a final appeal from Johnson & Johnson and allowed a $2.1 billion baby powder cancer verdict against the company to stand.

Why am I sharing these examples? Definitely not to condemn chemical or pharmaceutical corporations or the scientists working for them! I used to make a good living working for some of these corporations. My goal is to demonstrate that upsetting naturally-occurring mass, momentum, and energy balances in our ecosystem and denaturing materials and life forms is often associated with large costs. There is a reason many processed and human-engineered products such as denatured fine particles are suspected carcinogens. Our species and ancestors did not face these denatured or concentrated forms of material so our physiology and anatomy, not having naturally adapted to these selection pressures, will not know what to do with these denatured products. We may survive some and die or mutate from others. A prime example of a popular denatured product is alcohol. Our body really does not know what to do with alcohol concentrations in excess of those found in moderately fermented natural food. Initially, our body reacts to alcohol with vasodilation (blood vessels relax) and a neurologically inhibitive response (elation), but later with vasoconstriction (shrinking of vessels) and a neurologically sympathetic (anxious) response that leads to hangover and headaches. Alcohol is not available in nature at high concentrations except as a toxin to stem and progenitor cells (i.e. life).

Besides cost and health considerations, the other major problem with denatured materials is the difficulty in stabilizing their mixtures. Materials in their natural forms have evolved over thousands of years into wholesome stable (balanced) products. But when through science and technology we ingeniously override forces naturally holding materials together we face serious challenges. For example, addition of anti-caking agents to any compound could upset the pH (acidity measure), viscosity (resistance to flow) or color of the mixture, which means it would require additives for stabilization and counter-balancing the effect of anti-caking agents. These new additives would in turn cause further changes in properties of the mixture, hence often requiring additional additives[iii]. This is why typical shelf-stable consumer products like shampoos and processed food like baked items often have a very long list of synthetic additives and ingredients. The next time you buy some processed food or personal care product, look to see if you can identify the anti-caking agents among the long list of synthetic ingredients on the label. Remember, a formulator's job is to achieve the product's end use properties while keeping the

---

[iii] That is why we call them *additives*, because they are often artificially needed *additions*

cost down, the appearance appealing, the ingredients uniform, stable and shelf-stable, and the process and packaging economical. That is a tall order and it is unfair to expect the formulators to worry about our blood sugar and cholesterol levels in addition to all that alchemy.

Since we are discussing normal particle sizes in nature, you may be aware that humans have now entered the realm of nanoparticles, i.e., particles that are about 100,000 times smaller in diameter than the thickness of a hair strand or a piece of paper. This is a realm invisible even to most microscopes. I believe these are still uncharted territories with respect to natural balances and human health. Although years of research and billions of dollars in funding has been spent on nano-scale technologies, new tissue studies indicate that engineered nanoparticles could damage DNA and lead to cancer[3]. According to a 2019 paper[4]: *"While most cancer nanomedicine is designed to eliminate cancer, the nanomaterial per se can lead to the formation of micrometer-sized gaps in the blood vessel endothelial walls... accelerating metastasis."* Again, as a materials scientist, I am not surprised by such discoveries because our natural evolutionary path did not select us to face these denatured nano-sized particles. For example, the gap junctions in our intestinal epithelium layer are larger than 2 nanometers, so not evolved to block a smaller size contaminant in our diet. Our skin pore size also did not evolve to handle particles smaller than 10-50 microns (Opening of our sweat or oil glands)[iv]. Our immune system has also not evolved to adapt against non-biologic nanoparticles the same way we adapted to bacterial (1 micron range) and viral (0.1 micron or 100 nanometer) particles.

What is the lesson from all this? Going against a fundamental natural balancing force (Van der Waals) is possible but energetically costly and a prelude to a number of possible unknown complications.

Olestra and Homogenized Milk

Solids and powders may be the preferred form in non-living nature but life forms mostly rely on aqueous (water-based) systems. Milk is one such product. Despite its fat content, milk is mostly water-based. There are natural colloidal forces in the liquid phase between small droplets, like those which cause coalescence of milk fat in raw milk into cream, the source of milk's beneficial fat-soluble vitamins A, D, E and K. Again, humans have tried to overcome these natural (balancing) forces through elaborate inventions. For example, when humans decided to live in large dense urban clusters (a denatured state by itself discussed further in chapter 5) milk from domesticated cows once supplied *fresh and raw* to local neighbors became a corporate-traded commodity that had to be transferred, cost-effectively, across long distances to large population centers. The commoditization of fresh milk was in response to, as

---

[iv] According to a study, Chinese women have some of the smallest and Brazilian women some of the largest facial skin pores, those visible to the naked eye. These patterns, noted carefully by formulators of cosmetics for women, are possibly induced by epigenetic and environmental evolutionary factors in different countries. Source: F. Flament et al., "Facial skin pores: a multiethnic study," *Clin Cosmet Investig Dermatol.* 2015; 8: 85–93

with other aspects of our new evolutionary path, the simultaneous need for more reach (distance), speed (from cow to factory to warehouse to store shelf to home), convenience (of supermarket vs. driving to a farm) and quantity (large population centers). Reach, speed, quantity and convenience form the holy grail of achievement by *Homo sapiens*.

To improve its shelf-life in transit and on stores' refrigerator shelves, raw milk was pasteurized at high temperatures and then homogenized. Apart from the fact that high temperature pasteurization is shown to denature milk proteins and destroy milk's beneficial vitamins[5] and enzymes[6], the process of homogenization further denatures the milk by breaking up, through force, large fat (cream) globules in the emulsion (liquid in liquid dispersion) from an average diameter of 5 microns to an average of about 0.5 microns. One micron, or micrometer, is one millionth of a meter (still 1000 times larger than a nanometer). This, as we learned from our fine particle examples, increases the number of fat (cream) globules by about 1000 times and exposes them to surface oxidation. It is reported that this process not only oxidizes and damages[7] conjugated linoleic acid (a valuable cancer fighting compound in raw milk) but also produces harmful cholesterol oxidation products which are atherogenic (causing formation of fatty plaques in the arteries), cytotoxic, mutagenic and carcinogenic[8]. In fact, Dr. Kurt Oster's milk studies from the early 1960s until the mid-1980s singled out milk homogenization as a major cause of the current epidemic of heart disease. Although Oster's conclusions are endorsed by Weston A. Price Foundation,[9] not every researcher agrees with them. Nevertheless, denaturing fresh milk to increase its shelf life as a mass-produced commodity seems to damage some of its key nutrients. As with other decisions we make in our new evolutionary path, we are trading off quality with quantity.

The opposite of breaking down naturally-occurring fat droplets is synthesizing unnaturally large fat molecules which are another form of denatured fat. This was exactly the corporate science project in the 1990s undertaken by consumer products giant Procter & Gamble (P&G). The company spent 25 years and $300 million to develop olestra, a synthetic (sucrose polyester) fat with polymer chains (unnaturally) too large to be digested through intestines as a nutrient. This was in response to huge market demands by people who wanted to eat fatty food and not gain weight by somehow forcing fat to bypass their natural feedback loops and fool their metabolism. So within 25 years, scientists at P&G wanted to outsmart (biohack) the human digestive metabolism which was at least 250,000 years in the making, evolutionarily speaking.

Despite protest by consumer advocacy groups about the side effects of Olestra, the FDA allowed the product to be marketed in the U.S. and the company spent large sums on TV ads, boasting about the use of product in Frito-Lay chips and calling the product marketed under brand name Olean as "one of the most thoroughly tested new food ingredients ever approved." On the advisory committee debating Olestra, 9 of the 17 members who ultimately voted in favor of the product, were food industry consultants, whom one member described as "acting as proponents" for olestra[10].

By making fat molecules too large and indigestible in the human digestive tract, P&G was hoping to revolutionize the food industry. People could eat any fatty food they wanted without worrying about cholesterol or cardiovascular problems. Here again, humans cleverly utilized science to shunt an essential feedback loop in the human body.

Unfortunately, humans who were anxious to shunt natural fat metabolism with Olestra paid a price for their trust in the product. FDA received over 20,000 consumer complaints about Olestra (most occurring in the first five years after its approval in 1996), more than about all other food additives combined in the history of the agency, according to Center for Science in the Public Interest[11]. The complaints included diarrhea, incontinence, and the leaking of a yellow-orange oil into toilet bowls and underwear. Researchers also found that Olestra decreased the absorption of important fat-soluble vitamins and carotenoids from food.

P&G corporate scientists may not have realized there is a reason our body naturally craves fat. In fact, a good bit of our energy is generated through beta oxidation in our mitochondria and the joint action of long chain fatty acids found in meat with acetyl carnitine, also found in meat. As described in chapter 4, eating too little fat may in fact trigger our body's "bury" mode, which means converting other nutrients to fat. That is one of the reasons behind the recent popularity of ketogenic (high fat) diets. A 2011 Purdue study also confirmed that eating fake fat like Olestra may actually lead to weight gain.

Health concerns kept olestra from receiving approval in Canada and much of Europe. The company slowly phased out Olestra from most products and started marketing and re-branding the product as Sefose in paints and industrial lubricants. Meanwhile, the quest to biohack the human's natural fat metabolism pathways continued in the 1990s by a host of consumer products and chemical companies like Arco Chemicals and Monsanto (now Bayer, manufacturer of chemical pesticides).

What can we learn from the history of Olestra and shelf-stable milk? That the natural intelligence of evolution has adapted our body, as other life forms, to actively and dynamically stay in a homeostatic balance with certain naturally-occurring products. As will be described in future chapters, this is mainly to remain, as a species, metabolically efficient and evolutionarily fit. Fat in our evolutionarily developed diet and natural form came in a size range that was small enough to be absorbed with other vitamins (and acetyl carnitine) through our digestive tract, but large enough not to penetrate through the arterial endothelium and cause atherosclerosis (buildup of plaques and fat on the arterial walls). This is the time-tested wisdom of life (biology) and evolution – natural balances and feedback loops, a million years in the making. After about half a century of fat-fighting science and products, new research now shows not all fat is created equal when it comes to heart disease. For example, the size of low-density lipoproteins (LDL or bad cholesterol) is an important risk factor in cardiovascular disease. Smaller LDL with particle sizes closer to 20 nanometers pose a higher risk for heart disease than larger LDL particles around 30 nanometers or larger[12].

****

It is not only humans that can be affected by denatured human-made materials. We can harm animals too. Recently we rescued one of our egg-laying chickens from choking on a piece of plastic (styrene-based) foam insulation she was trying to eat. Evolutionarily, chickens like other birds have adapted beaks to handle naturally occurring polymers (peptides) found in worms but not viscoelastic human-made polymers in plastic or elastic foams. On a different occasion, one of our chickens became very lethargic after we fed her a piece of bread made with high protein flour. She looked very sick and kept drinking water until she defecated which made us realize the high protein-to-fiber ratio in the flour made for humans is not natural and had caused her constipation. No wonder some high-protein low-fiber diets make humans constipated and lethargic. The lesson here is that denatured human-made materials can lead to imbalances and disease. An estimated one million seabirds die every year from plastic pollution[13] and who knows how many more die by ingesting denatured human food. Our symbiotic evolution has made our body compatible with a range of natural products, which we will invariably disturb with our engineered or biohacking products.

Before we start pointing a virtuous finger at corporations, let us remember that in a free market economy, businesses are just responding to consumer demands. When we need products that help us ignore natural, biological and evolutionary systems, businesses deliver that to us. This will be explored further in chapter 7.

The Wisdom of Nature: Colloidal and Gravity Forces

A basic premise of this book is that nature's wisdom is self-balancing across time and space. Life forms rely on self-balancing (intelligent) aqueous or aqueous/oil emulsions in nature. If you are wondering how nature handles these small particle dimensions and interparticle flocculation forces, you are wiser than I was when I attended graduate school. Although I specialized in colloidal science in my Ph.D. work, questions about the natural wisdom behind our scientific discoveries rarely came up for discussion and debate. We were mainly trained to get ready for industrial job opportunities. It was not until years later that I started to understand the wisdom behind cohesive forces between small particles and how nature handles this realm. The following are three biologically important examples that impact our lives:

1) During an injury or immune response to a pathogen, the smallest colloidal cells naturally stabilized (free floating) in our blood, i.e. platelets (about 3 microns in diameter) coagulate and may form blood

clots to patch up the site of injury and pathogen attack – the microscopic war zone between our white blood cells and pathogens. Of course, too much and too little clotting means imbalance in our body, and imbalance means disease, such as hemophilia (inability to clot upon bleeding) or deep vein thrombosis (clot formation caused by platelet aggregation).

2) At the same time our body uses microscopic platelets to plug excessive blood loss, it uses a combination of steric (spacing through long molecules), and electrostatic (ionic) stabilization mechanisms to protect our tissue cells from leaking, coalescing or contamination by what is in the blood. Because blood is water-based (hydrophilic or water loving) our cells use a double layer lipid membrane (fat-based or hydrophobic, water-hating) to protect their watery content. However, the lipid double layer has polar water-loving heads stretching externally into the water-based plasma (blood) and internally towards the water-based contents of the cell. These (phospholipid) bilayer barriers stabilize and protect the integrity of our cells. Human cells are on average around 100 microns in diameter and would coalesce and leak without these surface active bilayer barrier membranes.

3) Nature uses surfactants like casein (phosphoprotein in milk stabilizing fat droplets in water) or lecithin as cell membranes or for stabilizing small fat particles (like in egg yolks or our blood stream). Humans now use chemically extracted forms of lecithin (from egg yolks or from seeds) as an emulsifier and stabilizer in products like mayonnaise, infant formula, hand and body lotions, cosmetics and pharmaceuticals[v]. But popular inexpensive synthetic dietary emulsifiers such as carboxylmethylcellulose (CMC) and polysorbate 80 (P80) are shown[14] in animal studies to alter the composition of the intestinal microbiota and induce chronic low-grade inflammation, ultimately leading to metabolic dysregulations as well as social dysfunction and anxiety-like behaviors.

By applying the aforementioned natural concept of a polar-nonpolar layered membrane to stabilize colloidal size droplets or particles, humans have developed synthetic surface active agents (surfactants) such as soaps and detergents. The next time you buy a shampoo, look for compounds like sodium laureth sulfate (SLS) on the label. These are cheap surfactants whose job is to emulsify oil (in your hair or skin) into the washing water. The International Agency for Research on Cancer lists two of the contaminants in these surfactants as carcinogenic (ethylene oxide as a known human carcinogen and 1,4-dioxane as a possible human carcinogen)[15]. Other risks associated with

---

[v] The scientifically curious can check Guang Wang's Masters of Science thesis project at Iowa State University: "Functionality of egg yolk lecithin and protein and functionality enhancement of protein by controlled enzymatic hydrolysis"

overexposure to human-made (synthetic) surfactants like SLS is irritation and damage to eyes, skin, mouth, and lungs.

While colloidal forces reign in micron-scale realms invisible to our world, in larger scales and distances, forces such as gravity play a major balancing role in the universe. Although of a different origin, gravity is another example of forces exchanged between all objects in this universe, but unlike Van der Waals forces, gravity fields are often strongest around larger objects. The farther and the lighter the objects are, the weaker the force. This is the gist of Newton's Law of Universal Gravitation. For the purpose of this book, I call gravity another form of honest unfiltered balancing feedback (exchange of forces) by nature which we cannot easily block and from which we cannot easily escape. Like all other species on this planet we have evolved in the presence of earth's gravitational field so periods of high gravity (as in high g-force roller coasters or aviation) or low gravity (as in space missions) would disturb our natural balance with the planet. There are many active research projects to create artificial (rotational) gravity in space to minimize the harmful impacts of low gravity on astronauts. I believe due to a mismatch with our evolutionary roots, these technologies have a long way to go before achieving effects as portrayed in movies like The Martian or Star Trek. Here again, shunting natural balancing forces may be possible for humans but are also very costly.

Universal Conservation Laws, Navier-Stokes and the Butterfly Effect

Besides universal interatomic and gravitational forces, the next three most important laws in science that govern natural balances are conservation of mass, energy and momentum. Basically every object and organism exchanges with its ecosystem mass, energy and momentum (directional force) in such a way that the total mass, energy and momentum of the system plus its ecosystem remains constant. So the universe and life are all about give-and-take and trade-offs.

Let's apply conservation laws to our body as a transport system. First we define a unit of time, let's say 24 hours. Time is among the most important factors in nature as well as human-engineered systems because a steady-state balance relies on time-dependent transport phenomena such as heat or mass transfer. For example, if our body is hemorrhaging blood we only have minutes and not hours to return the system (our body) to homeostasis hence we could greatly benefit from medical interventions. But if instead of an *emergency*, we face *chronic* imbalances (diseases) accumulated in our body over time, healing and balancing our system through lifestyle and diet changes take time.

OK, let's apply some of these mass and energy balances to our body.

Conservation of Mass

Mass IN (such as ingested food through digestive tract and inhaled oxygen through respiratory tract) MINUS Mass OUT (such as urine through urinary tract, excrement and flatulence gases through digestive tract, gases such as carbon dioxide and water vapor through exhalation and some via sweating through skin pores) EQUALS Mass ACCUMULATION in our body

A typical 24-hour mass balance on an average adult could be[16]: 2.5 pounds of food intake + 8.2 pounds of water intake + 1.8 pounds of oxygen inhaled = 0.5 pounds of feces + 9.8 pounds of water urinated, exhaled, in feces or sweat + 2.2 pounds of carbon dioxide exhaled. Now in real life we slightly gain or lose weight on a daily basis so we will have some accumulation but generally speaking, for an adult human with a healthy metabolism, what goes in (pounds of mass as solid or fluid) will have to come out (pounds of mass as solid or fluid).

Mass balance applies to all beings. In the bovine world, for example, on a daily basis, an average dairy cow consumes about 120 pounds of wet feed and 300 pounds of water for a total of 420 pounds of solid intake plus oxygen inhaled as gas. She yields, in daily output, about 50 pounds of milk, 120 pounds of feces and urine, and more than 250 pounds of water vapor and carbon dioxide (and a small amount of methane gas) in sweat, exhalation, flatulence and burping[vi].

What is the significance of mass balance for humans? Well, as with other life forms, human bodies actively (intelligently and evolutionarily) self-balance so they will resist or bypass biohacks, imbalances and interventions. So, for example, diets low in magnesium or too high in calcium and water-binding molecules – like salt and sugar – could lead to water retention in the body and weight gain, and as a result of mass balance, less urine volume or less water in feces (dried up stool or constipated). That also means higher blood volume and possibly blood pressure. Diuretics (medicinal, beer, caffeine or natural herbs like celery and parsley), on the other hand, could increase our urine volume and help us lose water, blood pressure and weight, sometimes excessively and in an imbalanced way. So as you see, the conservation of mass is the universal balancing rule that has implications in how our body functions.

Our intake (food) and output (excrement) of mass also impact our ecosystems. Remember, conservation laws apply to all systems, subsystems and metasystems. For example, if we apply a mass balance not only to our body as a subsystem but to the larger system consisting of our home (including bodies in that household) plus our waste disposal (trash cans) and our sewage systems (excrements), the equations will show that our excessive consumption (input into the house and then our bodies as subsystem of that house) will certainly impact the accumulation in the total system (inside trash cans, septic tanks, our cupboards and refrigerators and our bodies and waistlines). If you run a mass balance on an even larger ecosystem including landfills, rivers and oceans, you

---

[vi] Blaming cows for global warming is mostly debunked. Unlike cars that consume carbon which was locked up in the ground for millions of years and send it on a one-way path to the atmosphere, cows basically recycle carbon by burping out the same greenhouse gases they consume as grass. The 2006 FAO (United Nations Food and Agriculture Organization) report blaming greenhouse gases mainly on livestock was full of errors and retracted later.

start to see what we call pollution or contamination is often rooted in imbalances caused by excessive consumption and output of individual subsystems (households) within that ecosystem. Remember, unlike humans, other naturally evolved species within the ecosystem self-balance symbiotically within the whole ecosystem. What upsets the balance, such as piling up of toxic pollution, is a result of humans achieving their homeostasis in ways that are not symbiotic with nature.

Homo sapiens like to bypass mass balances so we can self-indulge in food without gaining weight. But weight loss biohacks, pills and diets cannot override our body's evolutionary feedback loops and organ-level mass balances. So without frugal eating, hormonal and metabolic balance, and a diet balanced in electrolytes it may be hard to avoid unhealthy fluctuations in our weight and other biomarkers.

Conservation of Energy: Fox Chases Rabbit, Human Chases Tesla

The second law that applies to all natural exchanges is conservation of energy. Like mass balance, energy balance tells us "energy into a system" minus "energy out of the system" equals "energy either consumed by, or generated in, the system." Energy and work are scientifically interchangeable because one can lead to another. That is why scientists use units like calories to measure both energy and work. A calorie is the energy needed to raise the temperature of 1 gram of water by 1 °C but it is also equivalent to the work (energy) needed to lift one pound of weight by about one meter. Again, let's run an energy balance equation on the human body as our system.

Our body is a good example of a mass-energy exchange system which, like other dynamic biological systems, generates energy by metabolizing organic compounds in food[vii] with oxygen into energy, carbon dioxide and water vapor. Humans are fundamentally a combustion engine – we use carbohydrates and rocket engines use hydrocarbons, we use amino acids and rockets use ammonium perchlorate. I will later share how NASA and Space-X apply conservation and balance laws to design rocket launches.

So let's run our energy balance equations on the human body as a system. On an average day, an average 31-50 year old male consumes 2500 kilocalories and an average female in the same age group 2000 kilocalories for a moderate-activity lifestyle. This equals the energy input to our body (mostly the chemical energy in carbohydrates and fats in our food and in inhaled oxygen, and occasionally thermal energy in non-food inputs such as heat from a sauna) minus the energy output from our body (mostly the chemical energy of exhaled or excreted water and carbon dioxide). The energy consumed by our body (the system) is used up by all sorts of work such as walking, lifting, thinking (around 20% of our calories), even by chewing and digesting food. If not used up, the caloric input to our body will accumulate as stored condensed energy (fat or glycogen) in the body for future use.

[vii] In chemistry and biology, organic usually means molecules containing carbon, oxygen, hydrogen, nitrogen

Because we metabolize mass into energy to do work, our body temperature is usually elevated above the environment[viii] but adjusted within a narrow range through complex feedback loops and thermostat-like control systems in a process called homeostasis (Homeo= same; Stasis= stable state) which parallels a design principle in chemical and industrial processes called steady state, which means that pressures, temperatures and flow rates (inlet, outlet) remain fairly constant[ix]. Homeostasis is evolution's gift to us that ensures, if left alone, that we are metabolically efficient and evolutionarily fit, balanced and adaptive.

When it comes to homeostasis and energy balance, however, humans are seriously handicapped. In nature, we rarely see animals gaining weight over time because natural evolution is all about energy efficiency so the conservation of mass and energy and the homeostatic process simultaneously ensure the input and output of energy and mass in a biological system are more or less in equilibrium and balanced over time. In humans, however, because of our particular evolution (detailed in chapters 4 and 5) we can frequently lose homeostasis and enter a state of allostasis (Allo= variable; Stasis= stable state) leading to weight and energy (mood, drive) fluctuations. For example, the energy balance and homeostasis for a fox could look like this: (A) Fox sees rabbit; (B) Dopamine (neurotransmitter for goal-oriented pursuit and motivation) is released in his brain providing the energy authorization to motivate his body to chase the rabbit; (C) As fox starts to chase the rabbit, glucocorticoid (the hormone signaling release of energy and glucose in times of sustained energy need) is supplied to the fox's brain and muscles to ensure he has the mental and muscular energy for the chase.

The energy balance in the world of a fox means he burns up enough energy (through mediators such as dopamine and glucocorticoid) in his brain and muscles to successfully find, chase and hunt a rabbit, which as a meal will then provide just enough energy that would allow the fox to cycle some energy for future rabbit chases, and leave some for mating to pass on his genes. Energy conservation in nature revolves mostly around food, fight, and reproduction.

In humans, however, the energy balance is totally upended: (A) Human sees Tesla (car model) or a friend's nice car or girlfriend; (B) Dopamine is released in his brain providing the energy to motivate him to chase the Tesla, girlfriend or the job that could pave the way to the car or girlfriend; (C) As the human chases the stressful job to be able to afford the Tesla, glucocorticoid (hormone for delivery of energy and glucose in times of sustained energy need) is released in the human brain and muscles but unlike the fox chase, the human chase of a car, job or girlfriend does not use a whole lot of muscles or calories because our pursuits are mostly conceptual and jobs mostly cerebral.

So while the body is receiving immense amounts of glucocorticoid (stress) hormone, and our energy-consuming brain demands calorie-dense foods, our

---

viii Actually our blood temperature is slightly higher than our cellular temperature
ix Homeostasis in humans is managed cooperatively throughout the body but the main switch is in an area of the brain called the hypothalamus, as explained in chapter 4

body is actually not burning all the glucose released into the blood because there are no actual physical chases challenging our muscles and metabolism. This is different from the type of challenges other species and our hunter-gatherer and farmer ancestors faced. As we consume more calories than we burn, we gain weight. Also, the excess glucocorticoid (energy release) hormone in our brain and body leads to anxiety, depression and disease, often accompanied by elevated body temperatures and heat loss to make up for the low physical activity.

As we use biohacks or medical interventions to blunt the impact of allostasis and imbalance, we become even more metabolically inefficient. In fact, humans already eat and burn 27% more calories (per weight) than chimps and 50% more than orangutans, and we store a lot of it as fat. The average male *Homo sapiens* body has 12-20% fat vs. less than 0.1% in male chimps[17]. A study of human evolution shows we have increased body fat and decreased relative muscle mass as compared to other primates[18]. More will be shared on our metabolic efficiency and handicaps in later chapters.

Our body's metabolic wastefulness has energy implications for the planet too. Spanish researchers who analyzed the typical Spanish diet found that human excretion plus the energy needed for wastewater treatment make up 17% of the overall global warming emissions[19]. So our evolution into metabolically inefficient species has profound implications not only in our bodies and minds, but also in our septic tanks, sewer systems and climate change. Although we consume less calories (food) per pound of our weight than some other species like birds, we simply waste a large amount of food (mass) because we are thermodynamically not efficient.

Conservation of Momentum: May the Force Stay with You

The third universal conservation law used by scientists and engineers is conservation of momentum. In Newtonian mechanics, momentum is the product (multiplication) of the mass and velocity of an object. The first and second conservation laws apply to transfer (transport) of mass and energy in systems which could be moving or static, but momentum captures the force involved in the motion of a system. We can store energy and mass but not force which is exchanged in the form of momentum. Moving objects, large or small, even gas molecules, transfer energy and force to each other upon collision. And unlike mass and energy, the transport of force through impact and collision is "directional" so in physics, momentum is represented by a vector (a variable with both magnitude and direction).

The rate of change in momentum of any object (acceleration or deceleration) in unit time equals the force needed for that change. So, for example, a football (or soccer) player weighing 200 pounds (90 kilograms) running at a speed of 15 feet per second has a momentum (force of impact) of 3000 lb.ft/second (200 times 15). Now if the football player suddenly collides with another football player weighing the same and running at the same speed and exactly in an opposite direction (remember momentum and velocity are vectors with a direction) they will come to a complete stop and the sum of their momentum before and after collision is zero (3000 ft.lb/sec in opposite directions neutralize each other). If the exchange takes one second, that is about 3000 pounds of force on each of them that must be absorbed by their bodies, helmets, brains, feet, shoulder pads, shoe soles, as well as the ground and air surrounding them. High technology helmets, pads and shoe soles can dampen and absorb some of these shocks but not all. As a result, numerous footballers report CTE (chronic traumatic encephalopathy) a condition causing damaging inflammation of the brain, after repeated impacts and concussions in their career. That is a direct result of nature's momentum conservation laws in action.

The conservation of momentum is the foundation for Newton's third law of motion: For every action (momentum or force) in nature there is an equal and opposite reaction (opposite momentum)[20]. In fact, from the moment the universe was created through the hypothesized big bang, expansion of particles and life in the universe has been dictated by balances in opposite forces. Now we can surmise why many of the earliest human philosophies, such as Yin-Yang (balanced) dualism, Taoism concepts of Wu Wei (effortless balanced action) and ziran (naturalness and balance), as well as scripts such as the books of Tao te Ching by Lao-Tze and the ancient Book of I Ching, were based on "balance" as the core principle of life and universe.

The Butterfly Effect and Navier-Stokes Equations

An inescapable consequence of conservation laws is that our world is connected from one end to another, also known as the *butterfly effect* first noted by mathematician and meteorologist Edward Lorenz. He noted that complicated weather forecast models will yield totally different results in response to slight changes in initial (time) and boundary (space limit) conditions. For example the formation and path of a hurricane in Asia could be influenced by minor perturbations in atmospheric pressure a week earlier when a butterfly flapped its wings in Europe or America, a different space (and time). Although this is a metaphorical hyperbolic example, it does follow the conservation laws because as we discussed, mass, energy and momentum are not created, just exchanged in space and over time. Real examples include hurricanes in North America that are started as small atmospheric temperature perturbations weeks earlier across the ocean, or sand and dust storms nucleated in African Sahara and impacting European weather weeks later.

The kind of equations mathematicians use for meteorological forecast is based on Navier-Stokes equations – named after 19th century mathematician scientists Navier and Stokes – , which applies the conservation laws to small units of a dynamic fluid (atmospheric air, for example) and then integrates the behavior of small finite elements over a large continuous system like planetary atmosphere[x]. These are complex equations that predict the future pattern of a dynamic fluid subject to flow (mass transfer), force (momentum transfer such as through wind or planet rotation) and contraction or expansion through temperature changes (energy transfer). While in graduate school in the early 90's I used the era's powerful UNIX-based computers to solve simple Navier-Stokes equations to model and visualize fluid flow patterns in a small pond. Recently, during a winter's walk in a nearby nature preserve, I came across a frozen pond which had beautifully captured the flow patterns inside the pond right before freezing, the same kind of pattern that took me several weeks to model and visualize using supercomputers.

I believe Navier-Stokes equations and our ability to numerically solve them are among humanity's most impressive science and engineering accomplishments. Some 200 years ago, before the age of Navier-Stokes, predicting weather patterns with accuracy would be considered witchcraft. The equations are now fed into supercomputers by NASA and companies like Space-X to design complex space missions. For example, during lift-off, as solid fuel is burning and converting to hot gases, Navier-Stokes equations are applied to the propulsion system balancing mass and energy transferred during the combustion process and the momentum transferred (gravity-defying force) by gaseous combustion products exiting the nozzles.

Discovering the utility of Navier-Stokes to explain natural transport phenomena should teach us that everything is connected through energy, mass and force ensuring the universe will ultimately balance itself. Humans may try to bypass nature by simply altering the kinetics (rate and speed) of natural processes but in the long run it is the energy efficiency and thermodynamics (direction of energy transfer) that will push processes towards their ultimate state of balance.

Even on the subatomic (quantum) level, the universe seems connected, harmonized and balanced in mysterious ways. Einstein has demonstrated that even if we separate two entangled particles in the universe, their *quantum* states remain correlated regardless of how far or isolated they are, what he called in a 1947 letter to Max Born a '*Spooky Action at a Distance*'[xi] like ghosts communicating with each other.

The Laws of Thermodynamics

Navier-Stokes and conservation laws allow us to calculate the rates (speed) of energy transfer in chemical or biological systems but not the direction of transfer or the chances of it happening. Thermodynamics is the science of how energy is transported in the universe and in which direction. It is important because energy, as discussed earlier, is the main (trade) currency of movement

---

[x] The equations are solved subject to fixed initial space and time conditions, or in scientific terms, the boundary and initial conditions.
[xi] In German: Spukhafte Fernwirkungen

and life in nature. When we feel hot, our money, a tool invented as part of our new evolutionary path, may get us an air conditioner in a city to help us cool down. But in wild nature, as in a remote desert, without our invented tools like money and electricity, energy is the currency of life so it is the enthalpy (or latent heat) of evaporation from our body that will cool us down and keep us alive. Both in the design of an air conditioner[xii] and our natural perspiration in a desert or a sauna, the first and second laws of thermodynamics apply. Nature uses energy as its main currency of life and balance, and thermodynamics is the science behind the flow of this universal currency.

The first law of thermodynamics is what we already covered as "conservation of energy." It basically states that in an isolated system, energy cannot be created or destroyed; it can only be transferred from one form to another. I will share its importance in the coming chapters on how the natural evolution of the human brain and anatomy relied heavily on conservation of energy and how our departure from metabolic efficiency is seriously harming us.

The second law of thermodynamics is also of consequence to our evolutionary path because it involves the duality of chaos and order in the universe. The concept scientists have devised to quantify chaos is Entropy, which means "Towards Transformation" in Greek, and is the portion of a system's energy (Normalized or divided by absolute temperature) which is not available for useful mechanical work because it correlates with the degree of disorder or randomness of molecules or particles in the system. In other words, entropy correlates with the chaotic nature of a system which squanders useful energy due to disorder. This is akin to squandering the useful energy of a group of rowdy, young, and energetic spring breakers due to disorder, infighting, and at times destruction of life and property in the chaos. Now if the same group were orderly, organized and coordinated, they could channel and convert their combined energy to useful work such as building a shelter for foster children. A building, like life, is an ordered structure with controlled entropy. Then again, spring break is a break from order and too much academic structure!

According to the second law of thermodynamics, the direction of all energy and work transfer in the universe is such that the total entropy increases or at best remains the same. In other words, from the standpoint of energy efficiency, systems and life forms which are chaotic, imbalanced and under allostatic loads squander more of their energy in thermal (heat) form and less on productive physical work. Again, think about the group of rowdy and energetic kids. Or better yet, think about an unhealthy body under allostatic loads.

Relevance to Human Life and Psyche: Szent-Györgyi and Jung

What is the relevance of the second law of thermodynamics to our lives and this book? Well, the human body converts low-entropy food to high-entropy energy (body heat), water vapor, carbon dioxide, feces and urine. So the net effect of our existence is an increase in the overall entropy of the universe. So

---

xii Refrigeration cycles use another ingenious human invention called the Carnot cycle

like other animals, our lives contribute to the overall chaos and inefficiency in the universe. But thankfully we are paired on this planet with trees and green plants that neutralize some of the chaos through photosynthesis, converting chaotic molecules of carbon dioxide and water into ordered and long (polymeric) molecules called carbohydrates.

The second law of thermodynamics simply means that the farther we move from a state of orderly equilibrium (homeostasis), the less naturally energy-efficient we become. For example, we now know that chronic stress in humans, mediated by glucocorticoids (Chapter 4) places our body in a state of allostasis and imbalance, which could lead to depression, inflammation and higher body temperatures. This is a high entropy transition towards chaos, disease and disorder. The ultimate transition to chaos in life forms happens when they die and their body decomposes. At that point, one entropic cycle (circle) of life that started by organizing (lowering entropy of) chaotic molecules into an organized life form (baby) ends with an increase in the chaos and entropy of death. In fact, one can define life as the only force in nature that can defy the second law of thermodynamics and create order out of disorder.

In inanimate forms, an example of ultimate chaos is an explosive reaction in which the fast heat released by the reactants creates a positive (destructive) feedback loop which leads to an out-of-control chain reaction. Actually, peace activists could use the second law of thermodynamics to scientifically argue against bombs and weapons because explosives, made from molecules stabilized by using up human and natural resources and applying useful mechanical and chemical energy, irreversibly release a whole lot of thermal (non-useful destructive) energy upon explosion. This process converts useful human resources and natural chemical energy into high entropy, chaotic thermal energy and destruction.

One can argue that according to the second law of thermodynamics inanimate forms of nature approach equilibrium mostly through chaotic steps that increase the system's entropy. Scientifically, this is captured by Gibbs free energy, named after 19th century scientist Josiah Gibbs. Gibbs energy (G) is the potential maximum reversible work (efficiency) latent in a system, and calculated (at a constant temperature and pressure) as $G = H - TS$, where H is the potential (latent) energy (enthalpy) in a system, T is temperature and S is entropy, so TS (temperature times entropy) represents the non-useful energy of chaos in the process.

Gibbs or available energy is one of my favorite scientific concepts because it reveals to us that when we design a path of equilibrium and steady states (homeostasis) in chemical, physical and biological processes, by minimizing chaos, we minimize the loss in the Gibbs energy and loss of useful work potential in the system (and universe). In other words, by remaining close to a state of balance we are not only being metabolically efficient but also helping with the productivity and metabolic efficiency of the universe.

According to Albert Szent-Györgyi[21], the distinguished biochemist who isolated vitamin C and identified the metabolic process in living cells (the Krebs cycle), unlike inanimate forms, living organisms prefer reaching equilibria through small step balances, dynamic patterns and order as opposed to chaos. The intelligence of life forms lies in the establishment and transmission of these self-sustaining, self-balancing patterns and dynamic orders. As Szent-Györgyi remarked: *"The most basic rule of inanimate nature is that it tends toward equilibrium which is at the maximum of entropy and the minimum of free energy. As shown so delightfully by Schrodinger in his little book 'What is Life?', the main characteristic of life is that it tends to decrease its entropy. It also tends to increase its free energy."*

On the other hand, disease, disorder and destruction are associated with loss of Gibbs and available useful energy in this universe. Szent-Györgyi viewed living organisms as smart, self-healing liquid crystal type molecules in a quantum state, coordinating and sharing energy and life through electrons. He attributed cancer to an electronic (free radical) imbalance and disorder at the molecular level. This unorthodox view placed Szent-Györgyi, a Nobel Prize laureate, at odds with establishment scientists and cost him some research funding.

Another prominent scientist and polymath who applied the second law of thermodynamics to human life was psychoanalyst Carl Jung. He proposed that our various psychological disorders and tensions are a result of the entropic and chaotic flow of energy triggered by the imbalances in the energy levels of our conscious (ego) and unconscious (such as our complexes and our persona) psychic states.

One important note about the second law of thermodynamics is that it shows the direction and natural tendency of change but not the rate (kinetics) of it. For example, graphite, a form of carbon used in some pencils, is thermodynamically more preferred than diamond (also a form of carbon) but once the ordered diamond crystalline structure is formed (subject to some 730,000 pounds per square inch of pressure and 2,200 degrees Fahrenheit in temperature) it would only tend to transition to the more disorderly (thermodynamically preferred) graphite-like structure at extremely slow rates. Here again, order (diamond) is a lot more precious than chaos (graphite).

Steady State and Homeostasis: Feedback Control Loops

Steady states are often thermodynamically favorable states because the chaos and noise associated with transient states are energetically expensive in terms of efficiency and useful available Gibbs energy.

Industrial systems achieve steady states through a process similar to that used by (endothermic or warm-blooded) biological systems to achieve homeostasis, namely baselines, set points and feedback loop controls. For example, in an average human, cells function optimally between 98°F (37°C) and 100°F (37.8°C). This is the baseline and one of the set points for the homeostatic process in the human body. Similarly, in industrial systems, optimal temperature ranges are set points for feedback loop process control. Negative or self-correcting feedback loops are used in designing control processes for chemical, mechanical, electrical, and even economic systems. The parallels of biological and industrial feedback loops are shown in the diagram[22].

To design such controls, we first define our "system" as a group of elements like reactors, people, human organs, or even units of economy. We then identify flows and exchanges between elements of the system. For example, in our body, blood, lymphatic fluid and neurons permeate the system in between elements (organs). These are inputs and outputs in subsystems. Then we define our target baselines and safety windows for critical variables such as temperatures or pressures. To achieve these target set points, data on the target variables is measured in response to a stimulus by sensors on a continuous real-time basis and fed back to system controllers or regulators that measure the noise/error (difference between the measurement and target set point). Because we are dealing with a time process, the system monitors a baseline not a single set point in time. The control regulator will provide its own feedback to effectors in the system (such as sweat glands in humans) until the error approaches zero and the system is operating within the safety window. The word negative feedback means if the impact of the stimulus is too much, the feedback will be in the opposite direction to tamp down and self-correct the overshoot and excess.

Although rare, positive feedback loops exist and often lead to catastrophic outcomes. In reactive systems, an explosion results, for example, when the fast heat released in early reactions catalyzes (speeds up) reaction rates between unreacted reactants, which in turn release even more heat, and the positive self-feeding (destructive) loop continues and leads to an out-of-control chain reaction called explosion. In wild nature, rare chaotic events such as migration of invasive species into new habitats, or chaotic events such as volcano eruptions, tsunamis and earthquakes could locally lead to destructive positive feedback loops. Long-term, however, negative feedback loops are the norm because they maintain balance, continuity and global energy efficiency.

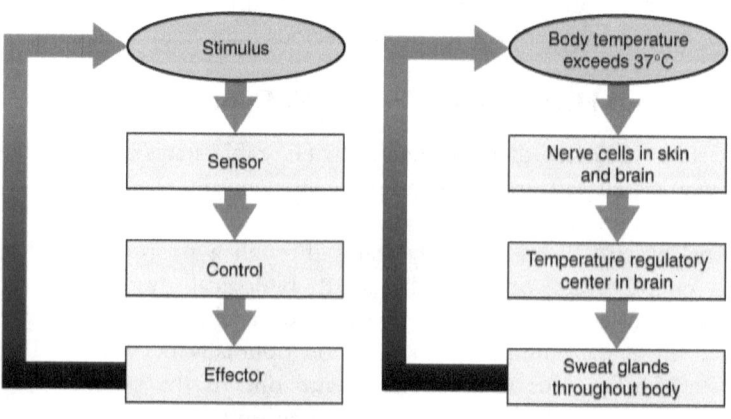

(a) Negative feedback loop   (b) Body temperature regulation

Here are a few examples of how negative feedback loops work to balance changes and noise in a process to keep it close to a steady state (equilibrium) baseline set point and hence efficiency. This is an important concept to understand when it comes to balancing any dynamic system, whether industrial or biological.

Central Heating or Cooling via Thermostats

Many of you may be familiar with this concept but it is worth reviewing. A central thermal regulation system in a house relies on sensor regulators called thermostats which sense temperature in zones (subsystems) in the house, measure errors (against set points) and if not zero (practically if not less than a sensitivity threshold like 0.5-1 degrees) then send a signal to the heating or cooling unit to supply or shut down the flow of heat or cold. This continues until the house zone temperatures get closer to the set points and the error, which is the difference between the set point and actual temperature measured, drops below the threshold. The feedback loop here consists of thermostat regulators and the furnace or the air conditioner.

This is similar to a mechanism our body uses for maintaining homeostasis. So it is important to know things that can go wrong in such a feedback loop control system:

1) If the house has thermal leakage (poor insulation) or an inefficient heating or cooling system, setting baselines at excessively high or low temperatures will not only run up your energy bill but will also gradually damage your heating or cooling system. In the human body, as I will discuss in later chapters, not all baselines are biological and involuntary. Due to our unique ability for abstraction and conceptualization, the psychological baselines we set in our mind (brain) often have enormous impacts on our biological baselines and can disturb our natural feedback loops. For example, it is shown that when depressed, our body temperature baseline (set point) is increased on average by about 0.5 degrees Fahrenheit[23]. Interestingly, one trick to reset this broken thermostat is whole body hyperthermia (heat treatment) which would kickstart body's balancing homeostatic circuits to fight heat-causing inflammation[24].

2) Sensors play a major role in feedback loops. I have visited houses with a single thermal zone that are freezing cold throughout the house both in the summer and winter even though the thermostat is set at 68 'F (20 degrees celsius). Upon inspection I found the thermostat to be right behind a 40" TV screen generating a lot of heat. This means the system's brain, i.e. the thermostat is unnaturally sensing an abnormally high temperature (generated by TV) which is not representative of the entire house. This is artificial zone control. In humans, this is akin to our biases and self-delusions, which provide us with a false sense of reality, thus upsetting our natural feedback loops from our brain to our

body. For example, the prosocial nature of humans means that when we attach to certain monolithic and like-minded organizations or political groups we often set our psychological baselines close to those of the group and experience group bias and stress in response to events that may not impact us individually at all. This could mean artificially-set baselines for our sympathetic nervous system and therefore, high blood pressure and other physiological disorders.

Natural Homeostasis, Feedback Loops and Metabolic Efficiency

As will be discussed in chapter 3, natural evolution relies on energetic parsimony or frugality and metabolic efficiency for long-term resilience. Homeostasis is a naturally evolved biological process that achieves metabolic efficiency through frugal use of energy (nutrient) resources. As a result, biological feedback loops responsible for homeostasis have these features in common:

1) They lead to energetic (metabolic) efficiency not only on an organism level but also on cellular scale, where homeostasis can be defined as a state where the production and consumption of metabolic resources balance each-other. For example, formation of ordered structure like protein macromolecules (i.e., life, enzymes, hormones via DNA), needs to be balanced by hydrolysis of Adenosine triphosphate (ATP), the molecular currency of intracellular energy transfer in our body. Humans have now developed medicines that target specific organs or cellular signaling pathways such as AMPK which control our cellular metabolism to induce recovery from chronic imbalances. But temporally blunting local feedback loops is analogous to artificial and dysfunctional heating and cooling zone control systems in buildings and homes and not the same as system or organism-level balance.

2) They rely on time and metabolic efficiency over time. That is why biological feedback loops often rely on time-dependent processes such as synthesis, release, circulation and binding of hormones in different organs. In our rush for speedy (artificial) recovery from disease or imbalance we trade off convenience and get-back-to-work speed with metabolic inefficiency (allostasis) in our body. An example of long-term metabolic efficiency is our brain's arborization (tree-like branching of neurons) and amazing ability for neuronal plasticity, to change and adapt as a result of learning and life experience. Neurons, through their synapses, impose energy demands over post-synaptic neurons in a closed loop-manner, modulating the dynamics of local circuits. This neuronal homeostasis over time ensures long-term metabolic efficiency by balancing ATP production (metabolism) and ATP consumption (synaptic activity)[25]. Researchers are now applying this Energy Homeostasis Principle to study and design learning in artificial neural networks. Brain's neuroplasticity will be explored further in chapter 4.

3) Biological systems are frugal and parsimonious. Chemical precursors, neurotransmitters, hormones and feedback loops in our body are often shared across multiple organs and systems. Our body achieves homeostasis by using signaling and messenger molecules (often proteins, also called peptides) such as neurotransmitters that carry (fast) neuronal signals in our nervous system, and hormones that are released and circulated in the blood. Since the goal is metabolic efficiency, humans have evolved in a frugal fashion so many of our hormones are derived from cholesterol and are precursors to, and balance, each other through feedback loops. For example, the male (fertility, sex and competition) hormone testosterone is biochemically almost identical to the female hormone estradiol and the two need to stay in a delicate balance to avoid problems such as infertility, low libido, insomnia, fatigue, weight gain, and mood swings. Our new evolutionary path, however, frequently upsets the balancing act of these hormones. Certain medications, stress and a large group of industrial chemicals called endocrine disrupting chemicals (EDC) have been shown to interfere with synthesis and homeostasis of steroid sex hormones and cause abnormalities in human sexuality, gender development, and reproductive capabilities.

4) Because biological feedback loops overlap and coordinate, organisms often rely on multiple signals and ratios of signals to self-adjust. When we consume food, for instance, our body has to metabolically decide to burn (calories), break down (fat and muscles), build (muscles) or bury (fat). It does so by relying on the balancing act of several hormones, which are messenger or signaling molecules released in the blood as part of this feedback loop. When it comes to energy metabolism, our hormones are usually either anabolic (using up glucose to build muscles or bury it as glycogen or stored fat) such as insulin, testosterone (androgenic steroid) and estrogen, or catabolic (breaking down fat, glycogen or muscle tissue) such as glucagon (often counterbalancing insulin), adrenaline and glucocorticoid (cortisol).

If you study the physiology of how the human body achieves homeostasis, you will find that it is almost always the ratios of different feedback signals that our body uses to balance itself in the homeostatic process. For example, our food digestion functions often get their cues from the ratio of fat to glucose to protein to fiber in our diet. Many other metabolic functions depend on the following ratios in the blood: sodium to potassium, calcium to magnesium to Vitamin D, zinc to copper, serotonin to dopamine, cortisol to testosterone, glucagon to insulin.

Even in trees, we now have discovered that the mitosis (cell division) and growth of vascular cambium is maintained by a network of interacting signal feedback loops that depend on the ratio of chemical and hormonal signals received from both the xylem (vascular tissue responsible for the distribution of water and minerals taken up by the roots) and phloem (vascular tissue responsible for transporting sugars made by leaves during photosynthesis to the rest of the plant). Interestingly, the growth signaling hormone in trees, auxin, has a molecular structure similar to the essential amino acid tryptophan[xiii] which the human body uses to make two hormones important for our homeostasis: serotonin – the patience, agreeableness, contentment and steadiness hormone – and melatonin – made from serotonin by our body's frugal hormone production factory and responsible for homeostasis of our sleep-wake cycles and circadian rhythm. When a tree's growth hormones and signal ratios are disrupted by genetic mutations through plasmids from a viral, fungal or bacterial infection, the tree isolates the abnormally rapid localized growth into a rounded outgrowth, called a burl. By doing so, trees, unlike humans, ensure the infectious cancerous growth does not disrupt the rest of the tree.

5) Biological baselines and set points are self-adjusted for long-term survival in the ecosystem so they are often set in competition as well as in symbiotic collaboration with other organisms in the ecosystem. Evolution relies on both competition and symbiosis. No species sets up its baselines alone. In fact, individualism and imbalance with the ecosystem, as will be shown later, is evolutionarily a maladaptive trait.

6) Unlike industrial systems, biological systems, when facing prolonged imbalances (called allostasis in recent years), resort to epigenetic changes or mutations to avoid death. If we metabolically abuse our body with excess calories and stress, for instance, our body implements epigenetic changes such as shrinkage of dendritic neurons in the brain to accommodate what it considers the systemic or chronic imbalance. Researchers now propose that certain disorders such as Parkinson's disease are due to the dying out of dopaminergic neurons because of their huge size, which makes them very expensive in energy terms[26]. Other examples of epigenetic adaptation are chronic allergies, immunosuppression, and downregulation of various cellular receptors in the body leading to disorders such as insulin resistance or glucocorticoid resistance.

Our Body's Homeostatic Metabolic Needs

So how does our body use feedback loops to achieve homeostasis? Let's focus on one of the most important feedback loops in our body for metabolizing glucose, needed for cellular energy needs and life. Let's say your body determines that your blood sugar (glucose) level to meet your body's midday metabolic needs is 100 milligrams per deciliter. This will be your midday

---

[xiii] New research shows tryptophan plays a crucial role in the maintenance of systemic homeostasis in our body because it integrates pathways in nutrient sensing, metabolic stress and immune response.

homeostatic target. So 2-3 hours after your lunch sensors (receptors) in organs such as hypothalamus (brain) and pancreas detect blood sugar levels and start, as part of a loop, exchanging feedback to bring your blood sugar close to the baseline target (set point) of 100. The pancreas, for instance, releases either insulin (when glucose levels are high) to expedite glucose uptake by cells, or glucagon (when glucose levels are too low) to signal release of glucose molecules into blood from liver, skeletal muscle cells and adipose (fatty) tissue.

As described earlier, hormones are our body's main signaling pathway to energetically balance itself. Metabolically speaking, hormones can be classified in two major types: (1) Anabolic hormones such as insulin, testosterone (androgenic steroid) and estrogen signal cells to take up glucose to burn or build muscles or glycogen; (2) Catabolic hormones such as glucagon (often counterbalancing insulin), adrenaline and glucocorticoid (cortisol) signal cells to break down fat, glycogen or muscle tissue to generate more glucose.

Upon receiving the proper metabolic hormonal signal, liver, fat and skeletal muscles, depending on their measurement of blood sugar and the ratio of insulin to glucagon, burn sugar, bury it (through glycogenesis or lipogenesis) or produce more of it (Gluconeogenesis, Glycolysis or Lipolysis[xiv]). This process control is coordinated through multiple feedback loops, as seen in the diagram,

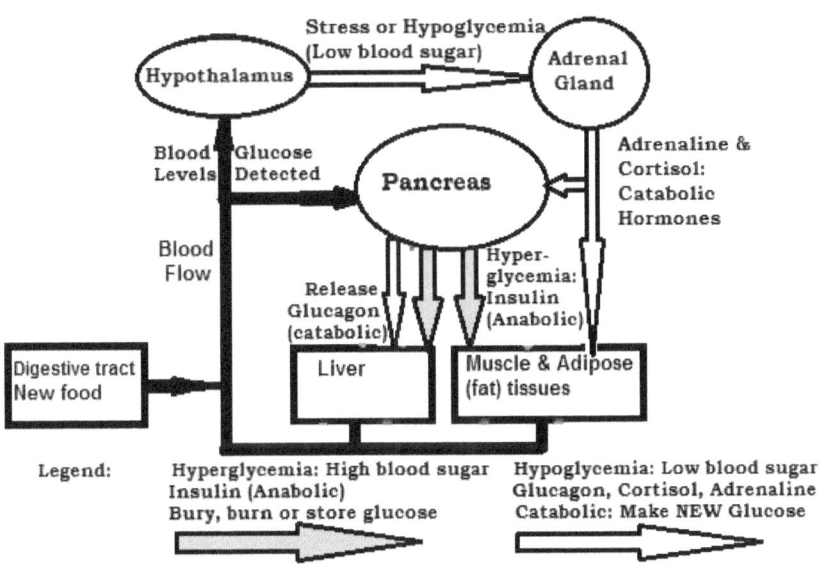

to maintain a blood sugar level of 100. Therefore, our organs obtain their cues from our diet (sugar in blood stream), metabolism and lifestyle (circadian rhythm) and do their best to keep us steady and balanced[27].

[xiv] Glycogenesis, lipogenesis, gluconeogenesis refer to formation of glycogen, fat cells and new glucose, respectively. Glycolysis, lipolysis refer to metabolizing (breaking down) of glucose and fat, respectively.

As will be discussed further, hormones are slow-acting signaling and feedback messengers. In times of emergency, such as a sudden attack or accidents, our body uses mechanisms such as electronic neural messages that travel fast down our central nervous system (CNS) and signal release of hormones such as adrenaline (from top of kidneys) or norepinephrine (released directly at target cells) that prompt cells for quick release of glucose. The point is that balancing a system as complex as our body is a time consuming process that is flexible and complex. Shortcuts or ignoring systemwide feedback loops can only lead to long term metabolic disease and allostasis.

Trees, Symbiosis and The Planet's Metabolic Needs

As we metabolize food in our bodies and hydrocarbons in our cars and industries, our planet's ecosystem needs to balance oxygen, carbon dioxide and energy levels in order to remain in a long term steady state. On a global level, variables such as average ocean temperatures are baselines and set points that need to remain more or less steady within seasonally safe windows to avoid catastrophic impacts on human civilization.

The natural processes involved in planetary homeostasis are obviously complex but photosynthesis and symbiosis are two important mechanisms involved. And perhaps the most important element in the planet's feedback loop are trees without which humans would perish. Trees are amazing because they simultaneously balance two feedback loops that are the most important to life on this planet: water and carbon dioxide cycles. For example, as the input of water (rain or snow precipitation) to segments of the planet exceeds the output (evaporation or streams into rivers and oceans), trees in that area sense the ground saturation and act as surge protectors or reserves for the water by increasing uptake. As the situation reverses and the grounds or the atmosphere dry up, trees release the water back into the environment to maintain homeostasis. This is why during hurricanes and flash floods, the majority of catastrophic water damage is incurred in areas that do not have many mature old growth trees. Trees act similarly to balance the carbon dioxide feedback loop by consuming carbon dioxide (photosynthesis) and converting it into carbohydrates and energy.

Interestingly, trees often do not act alone (selfish) and get their feedback cues for storing or releasing carbon or water from an underground (microscopic) network of mycorrhizal fungi filaments. This network connects tree roots together and transfers water, carbon, nitrogen, and other nutrients and minerals (phosphorus) to keep the overall forest ecosystem balanced and energy (nutrient) efficient through mutualism (symbiosis). Around 90% of land plants exchange some sort of feedback and are in mutually-beneficial symbiotic balanced relationships with fungi.

The mycorrhizal fungi filaments, in turn, remain balanced in mutualistic symbiosis with lichens (blue-green bacteria or algae) which provide fungi with nutrition through photosynthesis. Fungi are the crossing zone between animal and plant life, and lichens are nature's microscopic twilight zone where primitive bacterial prokaryotic life forms an overlap with eukaryotic life forms like fungi, trees and humans. If I could go back in time, I would dedicate my research career to these amazing undiscovered realms of nature. James Cameron's 2009 blockbuster movie Avatar depicted the electrochemical communication between the roots of trees on the forest moon, where the movie takes place, to remind us that all the organisms are connected so they can balance each other by communicating and collectively and efficiently managing resources.

Mutualistic symbiosis and balancing feedback loops are nature's main mechanisms for homeostasis and metabolic efficiency.

Unfortunately, most humans have parted ways from mutualism and other natural feedback loops. As will be described in the next few chapters, our lonely evolutionary path is metabolically and socially costly.

Chapter Synopsis and References:

Self-balancing is the universal wisdom behind life and nature. Inanimate objects remain balanced by exchanging feedback through natural forces. Life forms are governed by the universal conservation laws that dictate the distribution of material, energy and force in the universe, and the laws of thermodynamics which govern the direction and flow of energy. Natural evolution prefers equilibria, steady states and often mutualistic symbiosis because systems are most energy efficient and at peak available (Gibbs) useful energy when ordered and in balance. Nature uses feedback loops to maintain homeostatic and ecosystemic balance. Humans use feedback loop control systems to manage industrial processes. However, due to a unique brain and evolutionary path, humans can easily enter a state of metabolically-inefficient allostasis (unsteady balances or imbalances) and develop broken feedback loops.

---

[1] News archives: https://consumerist.com/2016/02/24/walmart-faces-lawsuit

[2] Figure source: Commons.wikimedia.org. 1939. File:The Canadian nurse (1939) (14584765199).jpg - Wikimedia Commons.

[3] Sara Pacheco et al., "DNA damaging effects of nanoparticles in breast cancer cells," 98th American Association for Cancer Research Annual Meeting, Apr 14-18, 2007; Los Angeles, CA

[4] Fei Peng et al. "Nanoparticles promote in vivo breast cancer cell intravasation and extravasation by inducing endothelial leakiness," Nat. Nanotechnol. 14, 279–286 (2019)

[5] Lauren E. MacDonald et al., "A systematic review and meta-analysis of the effects of pasteurization on milk vitamins, and evidence for raw milk consumption and other health-related outcomes," J Food Prot, 2011 Nov;74(11):1814-32

[6] Research summary by Robert Irons, Ph.D., "Pasteurization Does Harm Real Milk," published on https://www.realmilk.com/health/pasteurization-does-harm-real-milk/

[7] Carl Hans Norgauer, "Ecological study of the role of highly processed milk, meat and vegetable oil in prostate cancer causation," Universal Publishers, 2005

[8] H. Davoodi et al., "Effects of Milk and Milk Products Consumption on Cancer: A Review," Comprehensive Reviews in Food Science and Food Safety, April 2013

[9] https://www.westonaprice.org/health-topics/know-your-fats/milk-homogenization-heart-disease/

[10] Mayyasi, Alex. "The Fat Free Revolution That Gave America Diarrhea." Priceonomics, May 2014.

[11] Ballantyne, Coco. "Olestra Makes a Comeback--This Time in Paints and Lubricants, Not Potato Chips." Scientific American, 6 Apr. 2009.

[12] "Should You Seek Advanced Cholesterol Testing?" *Harvard Health*, 8 May 2014, https://www.health.harvard.edu/womens-health/should-you-seek-advanced-cholesterol-testing-.

[13] "Plastic Waste Kills up to 1 Million Sea Birds." The Ocean Conference, *Fact Sheet: Marine Pollution*, 2017.

[14] Holder, Mary K., et al. "Dietary Emulsifiers Consumption Alters Anxiety-like and Social-Related Behaviors in Mice in a Sex-Dependent Manner." *Scientific Reports*, vol. 9, no. 1, 2019, https://doi.org/10.1038/s41598-018-36890-3.

[15] "The Dirty Dozen: Sodium Laureth Sulfate." *David Suzuki Foundation*, 12 Feb. 2020, https://davidsuzuki.org/queen-of-green/dirty-dozen-sodium-laureth-sulfate/.

[16] "The Rubber Bag." *Https://Www.fourmilab.ch/*, https://www.fourmilab.ch/hackdiet/e4/rubberbag.html.

[17] Gibbons, Ann. "Why Humans Are the High-Energy Apes." *Science Magazine*, 6 May 2016, https://www.science.org/doi/10.1126/science.352.6286.639.

[18] Zihlman, Adrienne L., and Debra R. Bolter. "Body Composition in Pan Paniscus Compared with *Homo sapiens* Has Implications for Changes during Human Evolution." *PNAS*, 16 June 2015, https://www.pnas.org/content/pnas/112/24/7466.full.pdf.

[19] Muñoz, Ivan, et al. "Life Cycle Assessment of the Average Spanish Diet Including Human Excretion." *The International Journal of Life Cycle Assessment*, vol. 15, no. 8, 2010, pp. 794–805., https://doi.org/10.1007/s11367-010-0188-z.

[20] Mahajan, Sanjoy. *A Student's Guide to Newton's Laws of Motion.* Cambridge University Press, 2020.

[21] Szent-Györgyi Albert. *The Living State with Observations on Cancer.* Academic Press, 1972.

[22] "Category: Feedback Loops." *Wikimedia Commons*, https://upload.wikimedia.org/wikipedia/commons/2/22/105_Negative_Feedback_Loops.jpg. https://openstax.org/books/anatomy-and-physiology/pages/1-5-homeostasis

[23] Rausch, J.L., et al. "Depressed Patients Have Higher Body Temperature: 5-HT Transporter Long Promoter Region Effects." *Neuropsychobiology*, vol. 47, no. 3, 2003, pp. 120–127., https://doi.org/10.1159/000070579.

[24] Janssen, Clemens W, et al. "Whole-Body Hyperthermia for Treating Major Depressive Disorder." *JAMA Psychiatry*, JAMA Network, 1 Aug. 2016, pp. 789-795, https://jamanetwork.com/journals/jamapsychiatry/fullarticle/2521478.

[25] Vergara, Rodrigo C., et al. "The Energy Homeostasis Principle: Neuronal Energy Regulation Drives Local Network Dynamics Generating Behavior." *Frontiers in Computational Neuroscience*, vol. 13, 2019: 49., https://doi.org/10.3389/fncom.2019.00049.

[26] Pissadaki, Eleftheria Kyriaki, and J. Paul Bolam. "The Energy Cost of Action Potential Propagation in Dopamine Neurons: Clues in Susceptibility in Parkinson's Disease." *Frontiers*, Frontiers, 1 Jan. 1AD, https://www.frontiersin.org/articles/10.3389/fncom.2013.00013/full.

[27] Davis, Lawrence. "Efficiency of the Human Body." *Body Physics Motion to Metabolism*, Open Oregon Educational Resources, https://openoregon.pressbooks.pub/bodyphysics/chapter/human-metabolism/.

Chapter 3: Lichens, Cannibalistic Mating, SCOBY, Methuselah and Fibonacci: What Everyone Should Learn from Natural Evolution

Epigraph: Natural evolution is an intelligent process in that it ensures collective resilience and continuity of life from a gene-centered view, across species, time and space. This continuity is the result of a process which balances local chaos with long-range order (as in fractals), symbiotic with individualistic adaptation, within-group (competitive) with between-group (collaborative) selection. I provide examples from nature to demonstrate how this balance is achieved: Orangutans, cannibalistic maters, cheetahs, snails, lichens, SCOBY, the 4852 year-old Methuselah, sea slugs, sea nomads, Fractals and Turing natural patterns. I describe the mathematical algorithms and wisdom behind natural patterns such as arborization, which is shared by trees, our brain neurons and pulmonary bronchi. We will also learn that natural evolution rewards moderation, biodiversity and metabolic efficiency and punishes excess (greed) and monopoly (such as monocultures).

****

What came first, the chicken or the egg? Evolutionarily speaking, a dinosaur came first! According to research papers published in 2007, collagen extracted from bone fragments of a 68-million-year-old T. Rex Dinosaur closely matched the amino acid sequences of modern-day chickens[i]. So how does evolution work to convert over time the most ferocious creature on the planet to a most docile cute bird? And if natural evolution can convert dinosaurs to chickens, what does our past and future look like as *Homo sapiens*?

In chapter 1, I referred to some theories of evolution. There is nothing in theories of natural evolution which would be at odds with the belief in divine power among the faithful. The source of intelligence in natural evolution, such as life's origins, resilience and creation of order from disorder, can be assigned to cosmic events or a supreme intelligence (God among the faithful). This is not a book about religion or against it. Learning about evolution is as important as learning about our history. We study history in order to learn about patterns of past humans' sociological interactions and not to repeat their follies. We need to study our evolutionary roots in order to understand where, when and how we

---

[i] "Dinosaur Protein Preserved Over Time," by B. Prescott, *The Harvard Gazette*, April 30, 2009

biologically parted ways with other ancestral species (lineage), and how our new evolutionary path will impact our future.

Charles Darwin's widely accepted theory of evolution has these tenets:

(A) Life forms replicate/reproduce through encoded instructions we now call genes;

(B) Within the same species, the genetic code (genome) is not uniform and has variants, it also mutates over time;

(C) Because of the variants and mutations, species show a range of physical or behavioral traits. For example in frogs, certain variants in the genes that encode their leg anatomy result in stronger and longer legs;

(D) Those genetic variants encoding traits that can adapt better to selection pressures of their ecosystem win the arms wrestling over resources and reproduction, and will pass on to offspring. For example, gene variants that encode frogs with stronger legs ensure those frogs jump farther, survive and reproduce easier. As a result, those gene variants are selected and will survive along with those frogs to be passed on to the baby frogs which are likely to have strong legs and better than chance odds of survival;

(E) The surviving offspring continue the mutation, variation, selection and arms wrestling process and further improve their odds of selection (of their genes) for posterity.

So according to the new gene-centered view of evolution, it is neither chickens nor eggs but their software and the code (genes) that evolution cares about. Chickens and eggs are both used by nature as carriers (vehicles) and embodiments of genes. Nature uses these vehicles to test-run branches and versions of its genetic coding machine called life. The branches of life that are more efficient and outperform others are selected not by how the vehicle performed but by how the code ran and passed on its resilient genomic variants.

Think about different vehicle models, and styles or features. In the long-term, the manufacturers could not care less about the specific aspects or names of any models if they did not outperform other cars or their predecessor models (in terms of cost-performance balance, which is in turn a function of car design and design code). So the intellectual property that helps a manufacturer evolve its models and stay competitive is the design code of the vehicles. Nature's wisdom (intellectual property, if you wish) in evolving life is improving genetic design. And in evolution, the cost-performance balance is the metabolic (energetic) efficiency of each species in survival, reproduction and resilience in its ecosystem. So here we are again, back to energy conservation, balance and efficient feedback loops.

In his book *The Selfish Gene*, biologist Richard Dawkins asserts that a genetic lineage evolves to maximize its overall (inclusive) fitness in the ecosystem, which correlates with the number of copies of its genes passed on through all vehicles (not an individual organism).

So according to Darwinian and neo-Darwinian (selfish-genes) theories, natural life and evolution consisted of three simple goals: 1) Competitively hunting or scavenging enough food to metabolize into energy for securing future food; 2) Competitively mating and reproducing to pass on the organism's genes; and for, 3) Competitively protecting self, turf (habitat) and offspring to stay alive and ensure more copies of the organism's genes passed on for posterity.

But in 1967, a young evolutionary biologist named Lynn Margulis proposed that evolution is as much about collaborative symbiosis and successful mergers as it is about Darwinian competition. Like many visionary outliers, she was opposed by establishment scientists for years. Her theoretical paper, which was rejected by practically all scientific journals, demonstrated how in the process of endosymbiosis, the mitochondria (our cellular batteries if you will) in advanced life forms (eukaryotes) were once free-living simple bacterial life forms (prokaryotes) that somehow survived digestion (phagocytosis or endocytosis) by a primitive host cell. Simply put, our ancestral progenitor cells became what they ate! So through this mutualistic symbiosis, the new multicellular (eukaryote) organism became metabolically efficient and evolutionary adaptive. Margulis' theory was experimentally validated in 1978. She argues that random mutation, long postulated as the main source of genetic variation, is of only marginal importance. Much more significant is the acquisition of new genomes by symbiotic mergers and collaboration in nature.

So which path prevails in nature: The neo-Darwinian (Dawkins) or the endosymbiotic (Margulis) path? In my view, evolutionary fitness and survival of species depends on juggling life's crucial tasks (nutrition, reproduction, protection) in the most metabolically energy efficient pathway, whether via competition, collaboration (symbiosis) or a balanced combination of the two, which is often determined by selection pressures in the ecosystem.

Obviously natural evolution follows the conservation and thermodynamics laws discussed earlier. The following are some astonishing features of natural evolution. I have admittedly cherry picked and oversimplified these features to contrast them with features of the human's denatured evolutionary path, which I will discuss in chapter 5.

Cannibalistic Mating! Selfish Genes are Adaptive; Selfish Organisms are Not

Natural evolution favors species and gene variants that can pass on more copies. According to the selfish gene theory, whenever the interests of the vehicle (organism) and replicator (selfish gene) are in conflict, the gene wins and survives. For example, in certain spider species such as black widows and redbacks, the large females will often devour the smaller males during sex and mating – hence the widow in their names. Male spiders are instinctively attracted to cannibalistic mating despite the risk of death because the expected (probabilistic) value of survival of the selfish-gene prevails over the cost to the individual organism. The male spiders are such good obedient vessels (vehicles) for the genes that they even mutilate their own genitals that would result in partially disabling female's copulatory organs, decreasing her chances of mating with other males and increasing the reproductive success of the male spider's genes. So the male spiders mutilate and then sacrifice themselves all instinctively to keep their genes alive and selected by nature.

The Nifty Orangutans: Nature Selects through Physical Traits

The race for selection by nature is won by species and organisms that develop physical or behavioral traits conducive to metabolic efficiency, which is evolution's cost-performance balance. Nature's cost and currency is energy, and performance is survival, reproduction and resilience in the local habitat and ecosystem. An orangutan, for instance, would adapt to selection pressures in a forest habitat because its anatomy of long arms, toes and fingers plus a flexible biped– quadruped (two-feet – four-feet) posture have enabled it over the years to outperform many other species in metabolic efficiency. The adaptive physical traits and anatomy of orangutans help them be agile both on trees and on land for foraging and avoiding predators. An orangutan's arm-span is longer than its height and its hand has four long fingers plus a dramatically shorter opposable thumb for a strong grip on branches whether they climb the trees or hang from them. As a result, orangutans have adapted to become metabolically efficient in their natural niche in the ecosystem.

Since physical prowess is adaptive, nature discourages physical trauma. And as natural evolution discourages handicaps, we would be hard pressed to find many orangutans in nature that are missing a limb or fingers. When injured, animals must lay low and slow down to heal. If lucky and healthy, they will heal fast and be back on their feet before they become food for predators or decomposing microbes (if they die of weakness).

Stunning examples of physical adaptations can be found in deep sea creatures. When it is all dark, adaptation means you need to see very well but not be seen. These are exactly traits of a telescope octopus which is almost completely transparent and has protruding rotating telescopic eyes with a wide peripheral vision that enables the species to see both prey and predators with ease. The species is not only transparent but also does not cast shadows, which makes it almost impossible to detect by predators and even humans.

Snails and Cheetahs: Nature Strikes a Balance between Quantity and Quality, Speed and Resilience

Being faster, stronger and bigger does not guarantee fitness in nature. There is always a trade-off and balance in natural evolution between quantity and quality; between speed and resilience.

Being faster doesn't always guarantee winning in the gene-selection race. Snails and turtles clocking at 0.03 and 1 mile per hour, respectively, live as long or twice as long, respectively, as cheetahs clocking at 60 miles per hour. The protective shells of snails and turtles more than make up, adaptation-wise, for their slow speed by keeping them safe against environmental elements. Also, slow means slow metabolism, which makes snails and turtles energy (metabolically) efficient, definitely an adaptive trait.

Evolution also strikes a balance between quality and quantity. Species that are exposed to harsh, hostile unstable environments such as those found in deep oceans where temperatures are extreme and light, oxygen and food are scarce, evolve adaptive reproductive strategies to overcome these harsh selection pressures. The two common strategies are fast living and procreation cycles, or asexual reproduction which ensures species copy a large number of their genes before they perish. But asexual reproduction (no need for male and female mating) which is common among single cell or simple organisms such as fungi, bacteria, worms, starfish and lichens has its own checks and balances. It trades fast and easy reproduction of a large number of gene copies (quantity) with shorter lives and less diversity, resilience and adaptive flexibility to environmental and habitat changes. So over time it trades off higher quantity of genes with their lower quality. In nature, quantity (short-term scalability) and quality (long-term resiliency) are often at odds with each other.

Natural selection is about balance, efficiency and long-term trade-offs for ensuring continuity of life. That may be one of the reasons why the size of an average American household dropped by 2, from an average of almost 5 in the early 20th century to 3 a century later. Convenience and easy life means better survivability of genes and less need for their quantity.

Night Owls and Early Birds: Adaptation is Moderation; Excess is Punished

Intra and interspecies competition ensures balanced distribution of traits and curbing of excess through trade-offs. The natural ecosystem was a rough and wild world in which our early ancestors, like other animals, strived to survive and procreate in a relentless game of cat and mouse with each other and with other species. Although rough, natural evolution can be globally viewed as equitable because it levels the playing fields such that species remain in shape and improve their gene resiliency for their particular ecosystem. It may sound cruel to us but natural evolution often imposed boundaries through brute force by punishing organisms which showed hubris and disrespected their physical limits. Some might label it instant karma.

Nature's feedback loops are often in real-time. Small birds do not have late night parties like humans because birds that chirp at night are quickly detected and become food for the night owl. It is survival of the fittest and owls have the upper hand at night over most other birds. In addition to their night vision, owls' round facial discs act like satellite dishes and collimate sound waves into their ears. Even if small birds could party at night, they could not afford to have human-like hangovers in the morning because in nature "the early bird gets the worms." Adaptation often means moderation.

Natural evolution is not about one species dominating or exploiting others. As a result, everyone's level of consumption, procreation and aggression was kept in check by the necessity to remain strong, nimble and resilient. Greedy members of any species would either run out of food or become overweight, unfit and food for others. In chapter 1, the classic example of a peacock's tail demonstrated how nature teaches peacocks a lesson about vanity and the risk of ostentatious plumage. Natural selection ensures many traits follow a more or less Gaussian distribution around a mean[ii]. In fact, simple mathematical models of gradual Darwinian evolution in continuous time and continuous trait space, due to intraspecific competition for common resources, have found an exact analytical solution that shows emergence of Gaussian distribution of the trait[1]. Nature always runs mathematical algorithms which are not short-sighted or rigged by the powerful.

From Bears to Hummingbirds: Metabolic Efficiency is a Must

Natural evolution ensures bottom-up energy efficiency across layers of trophic levels, otherwise known as the food chain. The food (energy) chain starts at the bottom (first) trophic level with plants, trees, fruits, shrubs and algae that produce energy from sunlight through photosynthesis. The second trophic levels are consumers of energy, usually herbivore animals such as rabbits, cows and sheep which would be food for secondary consumers, often carnivores.

In a properly evolved food chain, about 85-90% of the energy in each level is lost as heat through organisms' bodily functions, excretion[iii] or bacterial decomposition upon death, and the rest (10-15%) is stored in the organism which would be food and energy for the next trophic level. As a result, apex predators high in the trophic energy pyramid, while dominating the food chain, have fewer calories available to them than lower level species. You can view this as nature's justice, because it balances resources and adaptive features across species. Alpha predators are not only food-challenged but also more vulnerable to toxins and pathogens that accumulate in the fatty and muscle tissues of lower level primary consumers. So even alpha predators eventually succumb to hunger and disease, especially as they age.

Earlier, we discussed the remarkable metabolic efficiency of blackpoll warblers equivalent to a fuel efficiency of 720,000 miles per gallon in human transportation terms. We also discussed the adaptive anatomical features in

---

[ii] Except in times of sudden ecosystemic change and evolutionary bottlenecks, leading to what evolutionary scientists Stephen Jay Gould and Niles Eldredge call "punctuated" evolution or equilibrium
[iii] Healthy humans excrete around 5% of their energy intake via feces

orangutans which make them masters of metabolic efficiency. Orangutans burn about 35-40% fewer calories per day as compared to average humans of the same weight[2]. If we add the non-food calories (oil, gas or electricity) humans borrow from nature and consume in their households (for heating, cooling, cooking, etc.) orangutans use about 90-95% less energy than humans. We will revisit primates and humans again in chapter 5.

No matter how large or small the species, metabolic efficiency is the key to survival in nature. Many species handle extreme temperatures or periods of reduced food (energy) availability by temporarily self-adjusting their body temperature and metabolic energy consumption rates. Such natural strategies include hibernation (as in bees, snakes, and woodchucks), torpor (bears, raccoons, red-tailed hawks, hummingbirds and swifts) or estivation (crocodiles, crabs, mollusks). Some birds such as hummingbirds can drop their body temperature by as much as 50 degrees Fahrenheit (28 degree Celsius) during extreme inclement weather or periods of food shortage.

Lichens, SCOBY and Immune System: Symbiotic Biodiversity

Life, naturally evolved, is a continuum that spans diverse species and organisms that symbiotically transfer materials and energy, balance each other and communicate through feedback loops. Mutualism is often preferred over parasitic relationships because it is more conducive to metabolic efficiency for the overall (eco)system. Nature uses bees to pollinate flowers, birds to spread seeds and turtles to recycle nutrients in terrestrial and aquatic systems, thereby using them as bridges of life connecting trees, flowers, soil and water.

Life is as much of a symbiotic chain as it is a competitive food chain. Natural evolution is most efficient when relying on a biodiverse and symbiotic continuum of life across space and time. This ensures energy efficiency through locally-adjusted feedback loops, energy transfer and balance, co-evolution and energy (life) propagation across time, space and dimensions from nano to macro scales[iv]. New scientific findings show that food chains and trophic levels continue below the visible scale to microscopic realms. We now know the crucial role of microbiome (microscopic bacterial biome) in the metabolic health of animal species and the role of fungi in plant species. Around 90% of land plants exchange feedback and are in mutually-beneficial symbiotic balanced relationships with fungi.

Humans have just learned in recent years that trees receive cues from their extended ecosystem to form their own metabolic baselines. Trees decide to store or release carbon or water after receiving signals from an underground (microscopic) network of mycorrhizal fungi filaments that connect tree roots together and transfer water, carbon, nitrogen, and other nutrients and minerals (phosphorus). The mycorrhizal network keeps the overall forest ecosystem

---

[iv] In chapter 2 we discussed how even inanimate objects impact each other and exchange feedback through atomic forces

balanced and energy (nutrient) efficient through mutualism (symbiosis). Mycorrhizal fungi filaments in turn remain balanced in symbiosis with lichens (a fungus living in a symbiotic relationship with an alga or cyanobacterium) which provide fungi with nutrition through photosynthesis.

Unlike most other life forms, lichens are not a single organism. Lichens are great examples of mutualistic (or commensal) organisms. Fungi protect the photosynthesizing algae or cyanobacteria that in return feed the fungi, all living as one organism. Morphologically, some lichens resemble a simple neural network, with the cortex and UV-protective layers provided by the energy consuming fungus (higher brain). Remarkably, lichens act as a twilight zone and bridge between different realms of life. If fungi are the crossing zone between animal life and plant life, lichens are nature's microscopic twilight zone where eukaryotic plant life meets primitive bacterial prokaryotic life[v]. This is aligned with the symbiogenesis concepts proposed by Lynn Margulis. In fact, thanks to their mutualistic nature, small size and slow growth, lichens are pioneer species, which means they bring life to most life-less habitats, including places lacking soil or moisture such as bare rocks, dead trees and extreme environments such as high mountain elevations, frozen soil in arctic regions, and dry soil in deserts. We have yet to fully understand or appreciate these organisms, but we owe a lot to them for the continuity of life on this planet[3]. And yet we hurt them on any chance we get. Not much can kill lichens except (human-induced) industrial pollution because lichens are nature's frugal and resilient life engines relying on just clean air for their metabolism and for creating life.

Another astonishing example of eukaryotic-prokaryotic microbe-level symbiosis is a SCOBY (Symbiotic Culture of Bacteria and Yeast), which is formed after the fermentation process of lactic acid bacteria and yeast to form beneficial sour (fermented) human foods and beverages such as kefir, kombucha and kimchi. Symbiosis gives a SCOBY such an evolutionary advantage that it is almost impossible to kill a SCOBY in normal ambient temperature ranges, even when dehydrating it.

Another remarkable instance of symbiosis at the interface of plant and animal lives *is elysia chlorotica*, a green sea-slug that has become photosynthetic thanks to a long-term symbiotic relationship with its favorite source of nourishment, the intertidal algae[4]. By partially digesting the algae, the sea-slug stores the photosynthetic chloroplasts in its digestive tract for stimulation by the sun when needed. So when hungry, the sea-slugs can crawl to the shores and sunbathe for the rest of their life cycle. Thanks to the symbiogenetic relationship with the algae, the emerald green *elysia* evolves to mimic the function, shape and color of a leaf to seize and metabolize sunlight and increase its evolutionary fitness by becoming solar-powered.

---

[v] By some accounts, a form of lichens fits "manna," the word used in the Bible and Quran to describe edible substance God provided to feed the Israelites following the Exodus

Professor Sidney K. Pierce of University of South Florida who investigated the molecular biology of the intracellular symbiosis between the digestive cells of a sea slug and algal chloroplasts believes endogenous retroviruses may play a key role in synchronizing the life cycle of the slug population as well as providing the means by which algal genes have been moved into the slug DNA. On his university profile page, Dr. Pierce indicates: "Transfer of genes between multicellular organisms has never been demonstrated before, so these species of sea slug may be a very useful model system to understand how such an important phenomenon could occur."

Humans have also co-evolved with microbes and viruses although in recent years we have focused on eradicating many of them. There is much we do not know about the role of microorganisms in the evolution of life and our innate and adaptive immune systems. But scientists are now starting to see the critical importance of microbes in our evolution:

A) Our immune system relies on leukocytes (white blood cells) performing the basic evolutionary process of phagocytosis (engulfing and digesting foreign exogenous cells), which parallels the endosymbiosis seen in primitive single cell microbes.

B) As for viruses, by some accounts, some 8% of the human genome is made up of Human Endogenous Retroviruses (HERV) or symbiotic viruses. These Endogenous retroviruses (ERVs) which we share with many vertebrates have been with us for a long time. In fact, modern immunological theories now view infectious diseases as a natural outcome of exogenous viruses and pathogens upsetting the delicate balance (homeostasis) between our body's immune response and these symbiotic viruses. This view is aligned with the microzymian theory proposed by 19th century French scientist Antoine Béchamp whose book was banned by the Catholic Church. Contrary to Louis Pasteur's germ theory which blamed exogenous microbes for our health issues, Béchamp blamed diseases on an (unhealthy) imbalance inside the organism and between the organism and its ecosystem.

C) As for gut microbiome (the good bacteria), a recent study shows they play a major role as agonists (activators) of intestinal receptors that contribute to the whole body's homeostasis and therefore prevent diseases, tumors, allergies or viral infections[5].

D) Parasitic worms, once removed at all costs from the human body, are now used as a medical treatment for certain diseases involving overactive immune responses. Scientists are looking for a connection between the elimination of parasitic worms and the increase in autoimmune disorders and allergies such as hay-fever.

Symbiotic evolution extends all the way from microbes to planetary scales, as discussed earlier (the Butterfly effect). In 1974, Lynn Margulis together with James Lovelock (a chemist), proposed The Gaia Principle named after the primordial mythological Greek goddess who personified the Earth. It proposes that symbiotic coevolution spans across the whole planet such that living organisms form synergistic and self-regulating systems with each other and even with their inanimate inorganic ecosystem to form a planetary biosphere acting as a dynamic adaptive feedback control system, which maintains the Earth in homeostasis and perpetuates the conditions for life.

Irish Famine and Destruction of Old Growth Forests: Monocultures Kill Life

When humans or natural conditions lead to imbalance and loss of biodiversity, nature imposes a penalty. For example, the Irish Potato Famine of 1845 was an outcome of limited biodiversity. Ireland's potato crops, which had mainly reproduced through asexual reproduction, were all vulnerable when a potato-killing plague swept the island. As a result, almost all crops failed, one third of the population either starved to death or migrated to America, helping with the genetic (human) biodiversity of their new homeland.

Today, agricultural scientists acknowledge that the industrial farming practice of monoculture (planting one crop in large scales) requires extensive use of pesticides and herbicides and is both labor and energy intensive.

In fact, even before the age of industrial agriculture, most natural farms lacked biodiversity because they were created by clear cutting old-growth forests. Any forestry scientist and professional can tell us that trees in the few remaining old-growth forests across the world live a happy healthy slow life[vi]. Unlike young forests, old-growth forests store large amounts of carbon. So destruction of these forests, which are currently not protected by international treaties, releases large amounts of carbon as greenhouse gases into the earth's atmosphere. Old-growth forests also serve as a reservoir for species which cannot thrive or easily regenerate in younger forests. Finally, old-growth forests are instrumental in generating fresh breathable air and pure water, regeneration of nutrients, maintenance of soils, pest control by insectivorous bats and insects, micro— and macro-climate control, and the storage of a wide variety of living genes on the planet.

Humans have already removed so much biodiversity that many species have become artificially – not through the natural impact of their trophic level and habitat – extinct or on the verge of extinction. The evolutionary importance of maintaining biodiversity is such that biologist E.O. Wilson, through his Half-Earth project, is calling for setting aside 50% of the earth's surface for other species to thrive in as the only possible strategy to solve the extinction crisis. There are also organizations dedicated to restoring biodiversity such as *The Center for Biological Diversity*.

[vi] As judged by markers of biodiversity such as mixed age stand, pit-and-mound topography, multilayer canopies, intact soil, healthy diverse fungal-lichen ecosystems, balance of downed and standing wood, balance of shade-tolerant and sun-loving shade-intolerant tree species, diverse wildlife habitat.

The 4852 Year-Old Methuselah: Slowness Rewarded by Evolution

As discussed in chapter 2, the second law of thermodynamics shows us that the tendency in the universe towards chaos and disorder is balanced by the need for maximizing useful energy and metabolic efficiency. As a result, nature often takes its time and changes happen slowly to maintain precious equilibria. I already shared how the human endocrine (hormonal) system relies on slow-acting frugal hormones to maintain homeostasis.

The wisdom of slowness in evolution is apparent from the way life has evolved in the solar system thanks to the sun, as the slow burning (in evolutionary time scale) source of life. An example of a slow natural process on the planet is photosynthesis, which turns solar energy into usable energy stored as carbohydrates or hydrocarbons. The reverse process and feedback loop which balances the photosynthesis cycle is metabolizing carbohydrates by life forms, also a slow and often efficient process. Combustion, on the other hand, which is a higher temperature, faster and less energy efficient process to metabolize hydrocarbons, was rare in nature before the emergence of *Homo sapiens*.

As already discussed, we owe much of life's genetic biodiversity on the planet, fresh breathable air, pure water, climate and pest control to old-growth forests, which in turn owe their health and age to slow symbiotic evolution with organisms like lichens and fungi. In terms of lifespan, slow metabolism and growth could even be more rewarding for trees than for animals. A member of the bristlecone pine tree family (Pinus longaeva) is the oldest known life form on the planet thanks to its slow growth strategy. Methuselah, with a verified age of 4852 years, is the oldest bristlecone pine in the White Mountains of Inyo National Forest in California. The specific location of Methuselah is a closely guarded secret for the sake of its protection from humans.

The secret to longevity of bristlecone pines is their metabolic efficiency, slow growth and adaptation to harsh habitats. Their small height (often less than 10 meters or 30 feet) means metabolic frugality. Their small footprint means less exposure to fire and elements. Their ability to slow down and become dormant in dry years means metabolic efficiency. Their slow growth, as low as 1 inch in diameter in a century, leads to dense, highly resinous wood that is a formidable barrier to invasion by insects and bacteria[6].

Fibonacci, Fractals, Arborization of Our Brain and Lungs: Locally Driven Adaptation Leads to Bottom-Up Scale-Free Evolution

In recent centuries, mathematicians have noticed many geometric patterns in nature seem to have a regularity to them and recur in different scales and contexts across the planet. It looks like many natural structures and morphologies first chaotically evolved, then adapted under some localized small-scale selection pressures before their success was replicated in larger scales. This is aligned with the theme of Darwinian evolution which is reliance on molecular level variations (chaos, randomness), followed by selection pressures (order and design) and reproduction (replication) on a larger scale, spatially and temporally.

The Italian Renaissance prodigy in arts and science, Leonardo da Vinci, was among the early observers of the wisdom of natural patterns. He noted that branches at every stage of the tree growth follow a pattern: When put together, they are equal in thickness to the trunk that originated them. What da Vinci did not know was that many biological networks such as his own body's pulmonary bronchi, brain neurons & intestinal microvilli use arborization patterns similar to that of trees, mathematically following patterns such as the Fibonacci series. The series, inspired by Indian Sanskrit Vedic studies, starts with 0 and 1, then followed by numbers that are the sum of their two preceding numbers, so it would be 0,1,1,2,3,5,8,13,..., growing by the Golden Ratio of 1.62. In 1754, Charles Bonnet discovered that the spiral phyllotaxis of plants (arrangement of leaves on a plant stem) were frequently expressible in Fibonacci number series. In 1917, biologist mathematician D'Arcy Wentworth Thompson demonstrated that the Fibonacci sequence could also describe the spiral growth patterns of animal horns and molluscs' shells (Fermat's spiral). There are many other natural patterns that more or less follow the Fibonacci series such as florets, fruitlets and flowers in sunflowers, pineapples and the artichoke[vii].

But what does the Fibonacci series tell us about the local, bottom-up and metabolically efficient nature of evolution? There is still much we do not understand. We know mathematically and evolutionarily speaking, random walks and symmetric growths are the norms in nature, yet in certain cases it seems to be metabolically more efficient for nature to choose an asymmetric (spatially or temporally) cell division over random or symmetric growth. This is nature's way of not putting all its eggs (genes) in one basket. Where one cell (mature) continues to undergo division every cycle and the other cell (immature) has a lag time for maturation and division, the resulting asymmetric temporal characteristic is shown to produce Fibonacci type growth. In the case of tree branches or florets in sunflowers this ensures a near-ideal packing that minimizes blockage of sunlight and maximizes spatial exposure for photosynthesis and energy metabolism.

In 1952, mathematician Alan Turing described how numerous patterns in nature arise from a competition between biochemical reactions and material diffusion. Turing's reaction–diffusion theory of morphogenesis has been applied in mathematical biology to model embryonic vertebrate developments, fish skin pigmentation and patterns, and the formation of lymphatic vessels.

[vii] The series has been used to explain all types of phenomena, from the ancestral tree of male honey bees to the number of possible ancestors on the human X chromosome, and in planning poker and the Scrum software development methodology. There is even a quarterly publication just dedicated to the series: https://www.fq.math.ca/

Turing's conclusions, similar to Fibonacci series, are based on conservation and the thermodynamic laws applied "locally" by nature in the smallest scale when developing (evolving) new beings, whether life forms or inanimate objects (like snowflakes). Sustainable growth of any organism or object in nature depends on its metabolic efficiency scaled up from the smallest scale.

Other astonishing morphological patterns created by a bottom-up natural evolution are arborizations (as in trees, brain neurons), spirals (as in snail shells, tornados, hurricanes), meanders (water streams or animal foraging paths) and waves (oceans, viral contagions, human brain's gyri and sulci). Whether a tree branch or our brain's neural network evolves new dendrites, or a budding snowflake crystal nucleus grows new spikes, whatever works thermodynamically and with conservation laws in the smallest scale often works in larger scales too. As a result, many natural patterns are self-similar at different scales. Whether you zoom in or zoom out, the patterns repeat themselves. This is the core concept behind another mathematical modeling technique called fractal analysis.

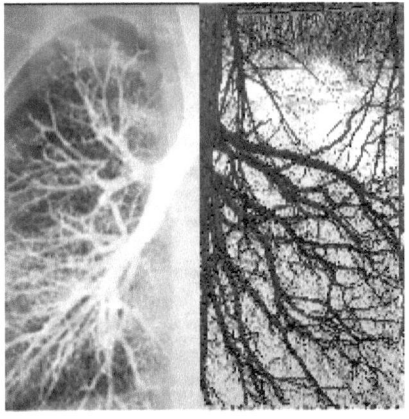

Fractals, first recognized by mathematician Benoit Mandelbrot in 1975, are self-similar scale-free patterns of structure such as those seen in snowflakes or ferns. If you zoom in with a microscope on the fractal pattern of a snowflake you get a nearly identical copy of the whole snowflake in a reduced size. Similarly, the fractal pattern of branching in a tree's twig is often the same pattern of branching that is observed on the tree's trunk. Mathematically,[7] the branching-within-branching structure of a fractal represents the best way to expose (pack) the most surface area (two-Dimensional geometry, such as a leaf or a cell membrane) within a three-dimensional space. For example, in the nautilus (shown here)[8], an ancient cephalopod mollusc some biologists call a living fossil, each chamber of its shell is an approximate copy of the next one, scaled by a constant factor and arranged in a logarithmic spiral.

Fractal-like patterns occur widely in nature, in phenomena as diverse as clouds, river networks, geologic fault lines, mountains, coastlines, animal coloration, snowflakes, crystals, blood vessel branching, tree branches and pulmonary bronchial tree (both shown in the image), brain neurons & intestinal microvilli and ocean waves. In fact, in the human brain, the fern-like structure that brings sensory and motor information to the cerebellum is called "the arbor vitae", which in Latin means the tree of life, because it is branched like a tree in a fractal-type fashion and is vital for accurate timing and coordination of our movements. What is the evolutionary wisdom behind the fractal-like arborization of the arbor vitae? As with other fractals, the arborization maximizes exposure of the myelinated nerve fibers used for relaying information or what biologist Bruce Lipton calls "cellular consciousness," in the small cerebellum volume. This can be credited for the cerebellum's massive signal-processing capability which is responsible for our body's movement coordination, motor learning and calibration of sensorimotor relationships.

Fractal patterns are even seen in animal behavior and interactions with their environments on spatial and temporal scales. For example, foraging paths by animals and hunter gatherer humans seem to follow a complex combination of chaos and order, best described by a fractal pattern[viii]. Because they are part of the species' evolution to achieve metabolic efficiency in their ecosystem, these patterns can be thought of as baselines for the species' homeostatic state of migration or movement in the ecosystem. Behavioral ecologists now use fractal analysis to monitor the health of wild and free-ranging animal populations in their natural habitats such as bottlenose dolphins in response to human disturbance and stressors.

In summary, what do fractal patterns in nature teach us? That evolution, even in complex organisms and seemingly chaotic patterns, acts wise by starting with simple local rules of metabolic efficiency, conservation and balance, followed by propagation of the rule (wisdom/pattern) over space and time. Start small, be patient, perfect the evolution, scale up. If you have not seen images of fractal patterns, I highly suggest you look online for stunning images and videos of natural fractals, Mandelbrot sets and Koch curves.

Rats, Worms and Sea Nomads: Evolution Has a Good Memory

As we saw with fractals, evolution copies successful patterns over both space and time. These are ecosystem-specific so environmental factors can alter heritable patterns. For the second half of the 20th century, it was a common scientific practice to assign heritability to human traits and deterministically attribute most diseases to ancestral genes. I still recall my heated debates in the 1980s with friends in the medical establishment. Unlike most of them, I was a believer not in the deterministic role of genes but in the strong impact of environmental factors on physical and behavioral changes over time. As a materials scientist, I knew inanimate molecules such as polymer chains, with molecular scale mobility, reacted over time to environmental factors. That is why many plastic products are prone to environmental stress cracking (ESC) and hence marked as disposable. There is even an equation (called Williams-Landel-Ferry named after its developers) that transposes polymer properties over time with environmental temperature. And because our DNA is nothing but pairs of polymeric strands I expected it would also be responsive to environmental factors.

[viii] Many ocean predators, birds and human foragers follow fractal paths consisting of random local walks interspersed by long trajectory forays, mathematically best described by a Lévy flight, named after mathematician Paul Lévy.

My views were vindicated in the 1990s, when the field of epigenetics gained attention and acknowledged that external or environmental factors impact phenotypes (heritable physical or behavioral characteristics)[ix] not through genetic changes but through impacting how genes are expressed. So time and environment do matter, and what we do today will impact our offspring, epigenetically.

We now know that even being caring as a trait can pass on to posterity through Transgenerational Epigenetic Inheritance (TEI). Studies in rats have shown that female rat pups born to lactating mothers, who engage in a high frequency of pup licking/grooming during the first week postpartum, become more caring towards their own pups in the future. Females born to low licking/grooming dams and then fostered to more caring dams will exhibit high levels of caring toward their own pups, which shows nurture and epigenetics play a role[9]. A more caring household often means a more caring offspring so the impact of nature and nurture on our evolution and development are not as independent or isolated as once thought. In humans, there is a striking trans-generational continuity in the case of child abuse. It is currently estimated that up to 70% of abusive parents were themselves abused.

Evolution has a long term memory, not only impacting our phenotypes (physical and behavioral traits) but also our organs. We also now know that the brain of intelligent species owes its evolution to their slow lifecycles (temporal) and large social circles (spatial). The modern views on the evolution of intelligence in large-brained vertebrates suggest that intelligence coevolved with slow life history (unlike in simple life forms) in response to socio ecological challenges when species form long-lasting social bonds. I will elaborate on this further in chapter 4 when I discuss the Dunbar's number, which refers to a cognitive limit to the number of members one can meaningfully relate to in a community. Human's wise brain and complex neocortex evolved for an optimal Dunbar's number of 150, higher than any other primate, thanks to our extensive social interactions, language and communication with other humans.

Our adaptive immune system is another astonishing example of how intelligence and memory is passed along through time by evolution. Our bodies are likely to have memories of many exogenous microbes that were immunogenic or pathogenic (disease– causing) to our ancestors. Otherwise, we would perish with the slightest scars, infections or a simple cold. If you are scientifically curious, I highly suggest you take an introductory course in immunology to learn about one of the most complex biological systems on the planet. Our adaptive immune system is so complex that even before the Covid-19 pandemic it was rare to find consensus among immunologists on the pathogenesis (origins and progression of infectious disease) for certain viral

---

[ix] In fact, my hunch was not off. Our DNA, composed of two polynucleotide polymer chains, like other macromolecules (polymers) is subject to configurational changes, hence properties, in response to environmental factors, through mechanisms such as methylation or acetylation.

infections and the *long-term* efficacy of artificial versus natural immunization (our own body's natural defense).

But intelligence, adaptive immune systems and Transgenerational Epigenetic Inheritance are not unique to complex organisms with slow life cycles. Recently, researchers at Princeton University demonstrated that even nematode worms (C. elegans, small simple organisms) develop adaptive immune systems and engage in transgenerational epigenetic inheritance. C. elegans not only develop an immune memory against pathogens by reading the small bacterial RNAs, but they also transmit this pathogen-avoidance memory for several generations through virus-like particles as epigenetic factors[10]. Remarkably, the adaptive immunocompetence memory can be transmitted both through germline cells and somatic cells (as crushed worm extracts exposed to naive worm cells).

What do we learn from all this? The wisdom of allowing time in learning, memory, intelligence and adaptation through interaction with the ecosystem. Natural evolution, uninterrupted and at its own pace, transfers through time and space the collective adaptive intelligence of life, whether it is metabolic balance and efficiency, predator and pathogen avoidance, optimum geospatial fractal patterns, or adaptive phenotypes.

As modern humans we are now trying to speed up and outperform the pace of natural evolution and adaptive memory through biohacks and technologies, yet early humans evolved naturally through a gradual adaptation process imposed by natural climatic selection pressures. For example, as *Homo erectus* moved out of forests, they gradually evolved[11] anatomical features such as a bipedal upright posture, shorter arms and fingers, less hairy bodies, more hairy heads (to avoid sunburn when standing) and smaller jaws, to better adapt to survival in the sunny open savannas and woodlands.

Thousands of years later, humans who migrated to islands or areas surrounded by water evolved other adaptive climatic anatomical features. An astonishing example is a group of people called "Sea nomads" who live, mostly on boats, among the islands and coastlines of Southeast Asia and collect their food by free diving, without SCUBA, oxygen tanks, wet suits or flippers, to depths of more than 70 meters (230 feet). Among the Bajau – a group of sea nomads who live on the waterways around the Philippines, Malaysia and Indonesia – divers often hold their breath for over five minutes while hunting for fish or shellfish. This is longer than most world-class free divers can hold their breath and about three times an average person's breathless underwater dive.

New anatomical and genetic studies[12] show that the heart rate of Bajau hunters during dives plummets to around 30 beats per minute, about 20 beats less per minute than the diving reflex of an average adult human. There are obviously adaptive epigenetic factors in play here. One anatomical genetic key to such stunning adaptation to the underwater environment is a variant of a gene that is associated with a large spleen, carried by about half of the sea nomads. A larger spleen means a larger reserve of oxygenated red blood cells available for underwater diving. When a human is submerged in water, the heart rate slows to conserve oxygen and the spleen constricts to expel a precious reserve of oxygenated red blood cells into the bloodstream. Some scientists have likened these evolutionary adaptations to the ones that have allowed Tibetans to thrive at high elevations.

So while natural evolution takes its time to epigenetically evolve a memory of resilience for species in a new ecosystem, how do modern human bodies handle long distance migration or travel, like my million-miles club coworker mentioned earlier? Obviously, natural epigenetic adaptation takes much longer than a few days or months to kick in so we often resort to biohacks and technologies to adapt to a life of transitions.

Ants, Super Chickens and Traveling Salesmen: Group Selection, Mutual Altruism and Swarm Intelligence

As discussed, symbiotic relationships can describe coevolution of adaptive traits in many species, and symbiotic mergers can even lead to acquisition of new genomes and traits according to the endosymbiotic theory by Lynn Margulis. But traits such as altruism and collaboration in social animals like humans and bees do not easily fit the Darwinian, selfish-gene or even endosymbiotic theories of evolution. While these theories focus on within-group individual selection and fitness, for altruism and mutual kindness to prevail it should offer a between-group selection advantage and an adaptive or advantageous tradeoff between individual convenience/resources and the group's fitness against other groups.

Among the many explanations of altruistic behavior is the theory of "kin selection" between related individuals as a subset of "inclusive fitness theory" first proposed by W. D. Hamilton in the early 1960s. The theory gives a selection criterion for evolution of social traits such as unselfish sacrifices when social behavior is costly to an individual organism's survival and reproduction. The criterion is that the reproductive benefit to relatives who carry the social trait, multiplied by their relatedness (the probability that they share the altruistic trait) exceeds the cost to the individual. For example, in a family with four children, the kin selection theory indicates that their father's altruism towards the children is adaptive if his sacrifices ensure survival of the four children. Each offspring inherits one-half his/her alleles from the father (a relatedness of 0.5) so the father's unselfish altruistic sacrifices (elimination of one gene set) would be evolutionarily rewarded because it saves and ensures replication of two gene sets (4*0.5). In recent years, biologist E.O. Wilson has proposed a multilevel selection theory including genes, cells, organism, and finally the group levels. The different levels function cohesively to maximize fitness, or reproductive success[x].

Regardless of what theory evolutionary biologists use to justify adaptive nature of non-selfish traits, altruism has undoubtedly resulted in some powerful human colonies at least at tribal and national levels. Even among non-humans, a brilliant delightful study by[13] biologist William Muir has shown that within-group only selection pressures lead to devastating results. Muir's study involved a chicken breeding selection experiment. He segregated chickens based upon their egg production. His original goal was to improve egg-laying productivity by selectively breeding the best egg producers (known as Super Chickens) generation after generation. So he would select top egg-laying hens and place them together, and repeat the process again by selecting the best egg-laying hens from different cohorts (sample groups) and placing them with other best egg-laying aggressive super chickens to compete again for best egg-laying. Think of this as a track and field tournament for egg-laying. He hypothesized that within-group competitive selection pressures would progressively develop a better egg-laying hen population, generation after generation. As the control group, he also grouped average non-aggressive egg-laying hens together and monitored their productivity over time.

The results were shocking. In the control group, the average-performing non-aggressive chickens were plump, well-feathered, healthy, and producing more eggs than they were at the start of the experiment. The competitively-bred super chickens group was decimated and in disarray. Most chickens were pecked to death by the top three super chickens on the pecking order. Ironically, the within-group only competitive selection pressures led to death, disarray and low (re)productivity.

Muir's experiment is cited repeatedly in recent years as strong evidence against competitive training and grooming of super performers in society, corporations and academia.

From a metabolic and energy conservation standpoint, group selection and altruism makes sense as collaborative foraging and trust building reduce the metabolic cost of survival for individual organisms. Biologist E.O. Wilson credits much of social grooming, collaborative foraging and group selection to the eusocial nature of *Homo sapiens*. Eusociality arises by the superiority of organized groups (colonies) over solitaires and cooperative pre-eusocial groups. Although rare, the few species that adhere to eusociality such as ants, wasps, bees and mole-rats, rank among the planet's most dominant. The biomass of ants alone composes more than half that of all insects, exceeding that of all terrestrial nonhuman vertebrates combined. Early hominins cooperated to rear their children while other members of the same group hunted and foraged.

---

[x] Over the years, numerous game theory and mathematical models have been developed to identify evolutionarily stable strategies in groups. One such model is the *tit-for-tat* strategy (be nice only if someone was already nice to you) which maximizes the group's welfare in a specialized case of prisoner's dilemma.

According to E.O. Wilson, those early cooperative and mutualistic traits, rooted partly in our eusocial nature, helped humans to dominate land vertebrates[xi].

Eusociality is adaptive because the evolutionary costs of individuals foregoing competitive foraging and reproduction are compensated by the greatly reduced metabolic costs of survival. Infighting and competition is metabolically costly as we saw in the super chicken experiment. Eusociality can be viewed as a metabolic trade-off between short-term individual fitness and long-term collective (genome-level) fitness. In some ant species, a queen that might live for only a few months if alone can live for 25 years or more as part of a colony, producing millions of offspring in the process.

And it is not only food and protection that eusocial species share. I believe the dominance of eusocial colonies over other species is strongly rooted in their sharing of intelligence as well as social structure and division of tasks. In the case of honeybees, for example, ethologists (who study roots of animal behaviors) now use the term "waggle dance" to describe the particular figure-8 pattern dance of honeybees to share information with other bees in the colony about the direction and distance to patches of flowers yielding nectar and pollen, to water sources, or to location of best new nest sites. Humans have even figured out the language code bees use in the waggle dance.

A more advanced example of collaborative intelligence sharing is the concept of "swarm intelligence" which imparts amazing "collective brain" powers to large flocks and colonies. How does it work? It is basically a process of registering and concentrating intelligence over space and time. Individual group members, spatially distributed, each collect local and path-dependent intelligence, and share that with all others as a message signal, as neurons do in a neural network. The process leads to a distributed intelligence network that benefits the entire flock or colony[xii].

The concept was introduced separately by Marco Dorigo and Gerardo Beni trying to model metaheuristic artificial intelligence cellular networks after the foraging behavior of ants. Each individual ant, while away from the nest and searching for food, lays down a pheromone trail that the other ants can follow. If the early ant reaches food, she will return to the nest using the same path, leaving more pheromone on the way back. But if she fails to find food, she will not lay down any pheromone signal on the way back to the nest. Because pheromones have a half-life before evaporation, the pheromone markings on successful paths leading to food last longer. Now when later ants, through their own random walks (also called Brownian motion as in fluid molecules) or Lévy flights (as described earlier in a footnote) come across previously marked paths, they follow the trails and if they find food, add their own pheromone to the path hence further extending the signaling timeline. Later ants who do not come across these heavily marked successful paths, will meander to find their own

---

[xi] Although Wilson's views have been vigorously debated, his supporters believe phenomena such as suicide, male homosexuality and female menopause evolved through kin selection and eusociality.

[xii] We called this "sharing best practices" in my corporate days and spent a lot of time on it.

food and mark their own paths or just come back to the nest empty handed and without marking any path. This behavior, repeated by thousands of ants, will, over time, lead to a collective swarm intelligence of where food or water sources are in a decentralized self-organized system.

Examples of swarm intelligence in natural systems include group behaviors such as flocking, hunting or herding in ant and bee colonies, birds, deer and other animals. Other examples include bacteria growth, fish schooling, microbial intelligence, and by extension our own adaptive immune system (innate as well as memory B cells and T cells). Swarm intelligence algorithms are now widely used in metaheuristic artificial intelligence models (of mobile networks), multi-destination traveling salesman optimizations, Monte Carlo simulations, genetically modified organisms, military applications such as navigation of unmanned clusters of drones or vehicles, and even in crowd simulation CGI (Computer-generated imagery) software in movies like The Lord of the Rings.

Shared intelligence or networking is even observed in vegetative states. Thanks to a communication network of fungus roots (mycorrhiza) discussed earlier, white pines are known to coordinate with each other in a local ecosystem to mature and drop their female cones every 3-8 years to maximize cross-pollination and biodiversity.

What does all of this teach us in the context of this book? That adaptive traits and intelligence at the local level of an individual organism could translate into collective level resilience and metabolic efficiency. We can form strong dominant supercolonies when we temper our within-group competitive selection pressures (individualism) and balance it with group selection traits such as altruism and sharing of food, shelter and intelligence.

Chapter Synopsis and References:

We reviewed the science and mathematics of natural evolution to prove it is an intelligent process in that it ensures collective resilience and continuity of life across species, time and space, i.e., ecosystems. This continuity is the result of a process which balances local chaos with long-range order (as in fractals), symbiotic with individualistic adaptation, within-group (competitive) with between-group (collaborative) selection. It rewards moderation, biodiversity and metabolic energy efficiency and punishes excess and monopolies (monocultures are basically genetic monopolization). It balances quantity with quality, and speed and scale with resilience. It endows even the smallest life forms awareness (cellular conscience) and an ability to share this intelligence through space, via mechanisms like swarm intelligence, and time, via mechanisms like transgenerational epigenetic inheritance. It is slow and takes time but efficient in that it endows species with long-term heritable adaptive memories.

---

[1] Biktashev Vadim N., "A simple mathematical model of gradual Darwinian evolution: Emergence of a Gaussian trait distribution in adaptation along a fitness gradient," *Journal of Mathematical Biology*, volume 68 (2014), pp1225–1248.

[2] Pontzer, Herman et al., "Metabolic acceleration and the evolution of human brain size and life history," *Nature*. May 2016, 533(7603), pp 390–392.

[3] Lichens are such an amazing organism that there are organizations dedicated to studying and protecting them such as The British Lichen Society: https://www.britishlichensociety.org.uk/

[4] Saracino, Vanina. "A Viable Planetary Future beyond Extraction, Predation & Production." *Strelka Mag*, June 2020.

[5] Taniguchi et al., "Mechanism for Maintaining Homeostasis in the Immune System of the Intestine," *Anticancer Research*, November 2009.

[6] Wohlleben, Peter. *Hidden Life of Trees: What They Feel, How They Communicate – Discoveries from a Secret World.* Greystone Books, 2016.

[7] Allman, William F. "Mathematics of Human Life," *US News and World Report*, June 1993, pp. 84-85.

[8] Conti, Roberta. "Category: Nautilus." *Wikimedia Commons*, 30 Apr. 2015, https://commons.wikimedia.org/wiki/Category:Nautilus.

[9] Champagne, Frances A. "Epigenetic Mechanisms and the Transgenerational Effects of Maternal Care." *Front Neuroendocrinol*, June 2008, pp. 386–397.

[10] Moore, R S., et al. "The Role of the Cer1 Transposon in Horizontal Transfer of Transgenerational Memory." *Cell*, 2 Sept. 2021, pp. 4697–4712.

[11] Lieberman, Daniel. *The Story of The Human Body: Evolution, Health, and Disease.* Vintage Books, 2014.

[12] Llardo, M. A., et al. "Physiological and Genetic Adaptations to Diving in Sea Nomads." *Cell*, April 2018.

[13] Muir, William M., and Heng W. Cheng. "Chapter 9/Genetic Influences on the Behavior of Chickens Associated with Welfare and Productivity." *Genetics and the Behavior of Domestic Animals*, 2nd ed., Academic Press, Amsterdam, 2014, pp. 317–359.

Chapter 4: The Chemical Soup that Controls Human Brains: What Everyone Should Know about the World's Largest Battleground

Epigraph: I summarize the neuroscience and natural chemicals behind human motivations, habits and behaviors. Human brains originally evolved to budget the energy needed for the so-called 4 F's of survival: Food, Fight, Flight and F... (reproductive sex). However, when human brains developed an ability to imagine and conceptualize, they bifurcated their host's (human's) evolutionary path from other species. This first became a blessing but then a curse because it singled out humans as the only species in which the brain was in the driver seat, not the body, and therefore, metabolic energy efficiency was no longer an adaptive or balancing trait. As a result, humans constantly create ecosystems that easily condition their own brains into self-reinforcing (broken) feedback loops which lead to behavioral ruts, self-delusions and addictions into endless reward seeking, irrational fear (paranoia), recklessness, greed and narcissistic rivalry. This is an important chapter about our brain's inner workings so try not to skip too much even if parts of it become technical. It is more important than your phone or car manual, without which you cannot properly operate.

****

In 2019, four of the world's top entertainment companies - Comcast, Disney, Charter and ViacomCBS - generated a combined $260 billion in revenues, roughly the same as the total sales of America's bread, butter, dairy and meat industries. Our great grandparents who experienced world wars, hunger and economic depression, could not have imagined their great grandchildren would spend more on illusionary characters like Mickey Mouse, Spiderman and Harry Potter, than on their food. In a broader sense, the United States media and entertainment industry generates more than $700 billion in annual sales, more than the entire food sales in supermarkets, warehouse clubs and grocery stores. Digital entertainment corporations are now even copyrighting images of nature for profit. Recently, the video by British filmmaker Philip Bloom who filmed the Moon during sunset was blocked on social media by Universal Music Group, which claims copyright ownership to generic shots of the Moon[1]!

And this is just the entertainment industry. Once we include the entire gaming, digital information and social media industries, it becomes obvious that the dollar value of digital food produced for the human brain dwarfs the value of caloric food produced for the human body. We are a brain-driven species.

But the human brain needs a lot more than digital data for its maintenance. The global sales of drugs, both legal and illegal, to treat psychological disorders are increasing at an unprecedented rate. Mental disorders, particularly various forms of addiction, are on the rise, and not sparing anyone, even the wealthy and educated, as we shall explore later in this chapter.

How did the human brain become so needy and maintenance-intensive? When did images, data, virtual connections, illusionary characters and mind-calming drugs overtake the need for food in the human economy? How exactly did natural evolution select the humans' high-maintenance brains? Are playing games and watching videos adaptive traits for humans? Are there other animals who exchange labor (work and metabolic energy) for a pastime of fantasies?

As I will illustrate in this chapter, our interest in fantasy, as well as our vulnerability to psychological and emotional stress, are rooted in the human brain's unique ability for conceptualization, abstraction and imagination, the same features that empower our ingenious inventions but also our delusional deceptions, and insatiable addictions.

Recently famed comedian and social commentator Russell Brand celebrated his 18 years of sobriety from drugs and alcohol. That may sound trivial but neurologically, it means 18 years of *daily* dialogue, often battle, between his frontal cortex (the cognitive brain) and his dopaminergic limbic/sensual brain circuits that could evoke the addiction at a moment's notice. Not everyone is as lucky and determined as Brand[i]. Tragically, countless movie stars and celebrities, even many self-help influencers, are seriously challenged in their mental battle against addiction. By the end of this chapter, we will clearly see why addictive self-reinforcing feedback loops have become an innate feature of the human brain, and why human brains are home to endless battles, the outcome of which is manifested in the war and peace we see in the external world.[ii]

Our strengths and weaknesses as a species are all rooted in the unique evolution of our brains. I believe in digestible science so everything I share in this chapter is a simplified summary of discoveries made in brain neuroscience, evolutionary biology, sociology and behavioral psychology. I have provided references as footnotes for anyone curious about deeper nuances.

---

[i] Assuming that he has not replaced the old addictions with new kinds, like addictions to fame or fortune. Habitual displacements plague us as we kick one form of addiction for another.

[ii] In a related quote, Henry David Thoreau states in Walden: "*Be a Columbus to whole new continents and worlds within you, opening new channels, not of trade, but of thought. Every man is the lord of a realm beside which the earthly empire of the Czar is but a petty state.*"

How Did Evolution Rewire the Human Brain?

Like other species, humans evolved to be metabolically efficient in seeking and securing food, sex and shelter. There are two widely accepted hypotheses[2] for the evolution of a complex brain. According to the Ecological Intelligence Hypothesis, complex cognition is an adaptive trait when the species faces challenges in searching for food. According to The Social Intelligence Hypothesis, a larger brain is adaptive for eusocial species that rely on group living and maintaining complex and enduring social bonds, deception, cooperation, and social learning from conspecifics.

According to this latter theory, the evolutionary path that led to a larger more complex (cortical) brain for humans started with our tendency towards *eusociality*, i.e., the formation of colonies that relied on shared intelligence, resources and protection, as well as specialization of tasks among the members. As discussed in chapter 3, group selection, altruism and intelligence sharing through communication makes sense for eusocial species because collaborative foraging and trust building reduce the metabolic cost of survival for individual organisms in the colony. A model developed by biologists Martin A. Nowak, Corina E. Tarnita and Edward O. Wilson proposes that three distinct steps are essential for species to overcome eusociality's initial evolutionary cost[3]:

(A) First, species must form distinct groups and colonies in discrete locations some distance apart, motivated by search for food, nest or following flock leaders or parents.

(B) Second, species must build a defensible nest which houses offspring and accumulate cooperative adaptive traits that favor the switch to eusociality.

(C) Finally, individuals must develop genes supporting eusociality (quell the urge to roam from the colony), whether by mutation or recombination.

Many evolutionary scientists now believe early hominins such as *Homo erectus* evolved more or less as eusocial species. Members of a eusocial human colony had to communicate with others[iii], share intelligence, and develop individual expertise in specialized tasks such as foraging, scouting, hunting and protecting offspring. Division of labor meant developing a curious analytical brain to explore new realms, solve problems and connect dots.

As early hominins were evolving their problem solving (specialized, individualized) and communicative (collaborative) skills, they somehow were compelled to migrate from forestlands into open meadows and savannas. That meant a major shift of food (energy) sources and shelter which required anatomical and physiological adaptations, and better problem solving and communicative (language) skills which demanded increased brain capacity.

[iii] As discussed earlier, eusocial species use creative communicative signals to share intelligence such as the directional 'waggle dance" by honeybees. Interestingly, some ancient human languages are also "directional" such as the Sambali language, spoken in the Zambales province of the Philippines, which only uses directional reference words (mayanan, bagatan, baitan, libaba denoting the north, south, east, and west) for pinpointing the location of objects.

Once out of forest habitats, hominins no longer had access to tree-based shelters and food sources like fruits. The best known early hominin is a 4.4 million year old species from Ethiopia which, unlike primates, had anatomical features of walking bipedally (on two feet). Brain size in early hominins such as the Australopithecus was less than 515 cc (cubic centimeters), similar to chimpanzees and gorillas. Unlike their canines, molar teeth in Australopithecus were much larger than those of earlier hominins, and had thicker enamel, suggesting their diet included hard, low quality plant foods, roots and tubers that required powerful chewing for mastication. This gave hominids an advantage over wild chimps that spend six to seven hours a day chewing a fibrous diet which limits the amount of calories available for their physical and cerebral (brain) growth.

The genus *Homo erectus*, one of the early members of the genus Homo, lasted from 1.9 million years ago to approximately 100,000 years ago. As *Homo erectus* moved out of forests, to better adapt to survival in the sunny open savannas and woodlands and on a softer, less fibrous, more meaty diet, they gradually evolved[4] anatomical features such as: Shorter arms and fingers and a bipedal upright posture better suited for walking and running vs. climbing and swinging; Less hairy bodies, allowing a cool-down by sweating in warmer climates; More hairy heads, preventing sunburn when standing; and smaller jaws, suited for less chewing and more cutting.

Softer, meatier food meant fewer calories were spent on chewing and more made available for the body and brain growth. Also, as *Homo erectus* discovered the use of fire in cooking, they spent even less time chewing their food because fire tenderizes food and releases its nutrients and calories by a factor of two to three as compared to raw food. More caloric and nutrient availability meant a larger body size (versus chimps and bonobos) and a major increase in brain size (since the brain is an energy hungry organ), from around 500 cc in early hominins to around 1200 cc for *Homo erectus*[iv].

Thanks to the bipedal upright posture of the species, the particular anatomy of hands and their increasing availability meant the development and utilization of manual tools, such as spears and axes, to hunt and build shelters. So the bipedal posture, while anatomically rare and awkward in nature, and making us prone to spinal injury and back pains, was adaptive for our early ancestors in that it freed up our nifty and crafty hands.

Many evolutionary biologists today believe the large size and complexity of our body and brain have evolved as a result of these factors:

(A) Tool making and use,

(B) Selection pressures for social collaboration/communication, problem solving,

---

iv This is perhaps around the same time that vitamin D started playing a major role in the metabolism of calcium (in the presence of sunlight in open meadows) into bones and skeletal structure of the bipedal species, as well as our immune system.

(C) Cooked and soft diet that leaves more calories available for cerebral and physical activities and growth.

Around 250,000 years ago, the Neanderthals, which were rugged hominins with brains similar in size to ours, are believed to have populated colder climates in Europe. Fossil and DNA evidence suggest our ancestral species, *Homo sapiens*, evolved shortly after the Neanderthals but in the warmer climates of Africa, and later in Asia and Eurasia. The increased behavioral sophistication of *Homo sapiens*, as evidenced by large brain sizes (1400 cc), complex tool sets and clever hunting techniques, enabled them to dominate other species like the Neanderthals. Studies of ancient DNA extracted from Neanderthal fossils suggest although extinct, they occasionally interbred with *Homo sapiens* and many of us still carry some of their DNA[v].

As *Homo sapiens* became the dominant eusocial globetrotters, their brains evolved features not fully evolved in reptiles, and even in other mammals and primates:

(A) Specializing in tasks such as scouting, hunting, toolmaking, cooking, raising babies,

(B) Communicating with others to share (advanced swarm-type) intelligence through signals and languages, and eventually,

(C) Conceptualizing, abstracting and cognitively transcending through time and space.

Let's briefly review the three evolutionarily staged parts of the human brain, or the *Triune Brain* as proposed by neuroscientist Paul D. MacLean[5]:

(A) The *reptilian or lizard brain*, composed of the basal ganglia (striatum), cerebellum and brainstem (including the amazingly arborized arbor vitae discussed in chapter 3), controls *instinctual* vital bodily functions we share with reptiles, such as heart rate, breathing, body temperature and balance. The reptilian brain is evolved to meet our need for living, breathing, food and sex, as well as coordinated movements (cerebellum), habits and routines that we develop over time and require little thought.

(B) The *mammalian or limbic brain* emerged in the first mammals. It includes the hippocampus (Registering mood and contextual memory), the amygdala (Registering pain, risk, danger and uncertainty) and the hypothalamus (registering and homeostatically regulating body temperature and metabolic processes). The limbic brain is also called the emotional brain because it is responsible for *"affectual"* gut-feelings and value judgments that *construct* our emotions. For example, when we enter a dark alley covered with graffiti and trash and surrounded by unfriendly-looking loiterers, our hippocampus assigns a high contextual risk, and our amygdala senses high episodic risk and uncertainty. They

---

[v] The percentage of Neanderthal DNA in modern humans is around 2 percent in many people of European or Asian background.

project an alarmed state into our hypothalamus, which in turn will initiate a vigilant (sympathetic) response in our body and physiological changes such as a faster heartbeat. These changes are implemented by the autonomic nervous system and through hormonal action initiated in the pituitary gland.

(C) The neocortex or the *rational brain* prominently evolved in primates and culminated in humans. This is the part of our brain responsible for creative and cognitive skills mostly unique to humans such as the development of human language, concepts and culture, abstract thoughts, calculations, imagination, and consciousness. The cerebral cortex is divided into four lobes: The frontal lobe (the analytical executive decision maker receiving input from all parts), the parietal lobe (our spatial simulator based on somatosensory, visual and auditory input), the occipital lobe (our visual cortex and image simulator), and the temporal lobe (decoding and organizing auditory and visual signals into categories, language, speech, words and long-term memories). It is thanks to the neocortex that the human brain is adept in the 4 P's of risk management: Prediction, Prevention, Protection and Planning.

The size of the human brain's neocortex seems to correlate with the number of meaningful social interactions and connections early humans had in their eusocial circles. According to evolutionary psychologist Robin Dunbar who studied the social connections among groups of primates in the early 1990s, the size of eusocial groups is mainly influenced by the size of the members' brains. Because it takes brainpower to effectively collaborate, communicate, brainstorm (for problem solving) and compete with other intelligent colony members, Dunbar theorized that the size of the brain's neocortex must correlate to how many interactions a primate can handle. Dunbar then measured the human neocortex and estimated the upper limit of humans' ability to maintain meaningful social relationships at 150 contacts. This is called the Dunbar's number, which is a cognitive limit to the number of people we can meaningfully and intelligently relate to in a community. This is why the neocortex is sometimes called the social cortex. This is also why many early farming villages, migrant settlements and units in armies and corporate organizations have around 150 members.

Will the high number of contacts in the age of social media and internet increase the Dunbar's number and size of our brain cortex? It is too early to say, but the social cortex concept is about making meaningful, insightful connections and exchanges that require brain power and problem solving. Sharing emojis, cute or outrageous images, or cultish labels and thoughtless talking points, even in large quantities may not engage our cortex enough to adaptively enlarge it over time. In fact, we may even be regressing because new research shows, at least in animals, dependence on fancy gadgets and conformity to preset rules and boundaries leads to smaller neocortex (cognitive intelligence) than brains of animals that adapt to natural selection pressures in the wild. This makes a lot of sense in light of the principles of natural evolution which we reviewed in chapter 3. We cannot beat natural selection in selecting the wise and the resilient.

The unique triune brain structure in *Homo sapiens*, including a sizable cortical region, was so adaptive that human colonies expanded both in number and

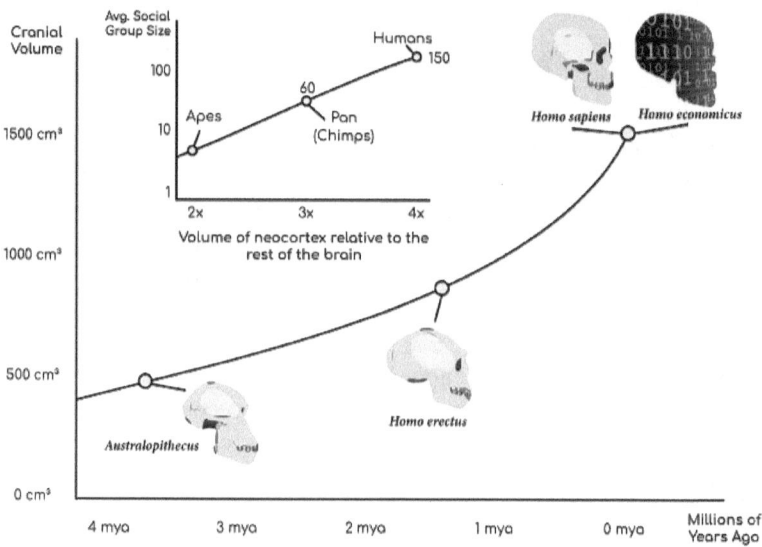

reach, from a small population in Africa to about 200 million around 2000 years ago spread across the globe, except in certain islands like New Zealand, Iceland, Bahamas, Hawaii and Madagascar which were mainly inhabited in the past two millennia.

More important than placing them on new continents, the humans' unique analytical, conceptual and emotional brains placed them on a new evolutionary path, which was grossly different from the natural evolution discussed in chapter 3. This was both a blessing and a curse. A blessing because it allowed humans to invent creative tools and rules to make life easier and to gain speed, quantity and reach over other species. A curse because it endowed humans with an unrivaled capacity for greed, deception, self-delusion, addiction, depression, anxiety and all sorts of diseases and disorders, like those discussed in chapter 1. Let us see why and how our unique brain separated our path from the rest of life forms.

Like other species, humans initially evolved to be metabolically (energetically) efficient in seeking and securing food, sex and safety, i.e., shelter and predator avoidance. Let's now try a thought experiment which best explains how our brain circuits evolved.

Imagine an early hominin - the general term used for extinct members of the early ancestral human lineage - still finding his way out of the forests and into open meadows. I will focus on the brain circuitry of a male hominin and refer to female physiology when needed. The brain circuitry of the hominin is evolved to find and secure food and sex safely and in a metabolically efficient way, which means spending the least amount of energy possible on the pursuit of rewards (sex or food). As our hominin is venturing out of the forestland and into new ecosystems, he should first *learn* about the *novel* features of the new terrain and develop some rules of thumb for assessing risks, rewards and survival. This is called the *"zone of proximal development"* in psychology, which is the learning zone, slightly outside the comfort zone (certainty) but not quite as far as the panic zone (high risk). So the concept of learning and building safety boundaries and baselines is always associated with rewards and risks (pain or punishment).

The same risk-reward calculation brain circuitry that allows dogs to be conditioned by trainers to learn to follow certain rewards and to avoid certain punishments, evolved in early hominins to learn the risks and rewards of their new ecosystems. This is the subject studied in the field of behavioral biology, also called *ethology*, which looks for biological root causes and mechanisms of our behavior. An early influencer of the field was physiologist and neuroscientist Ivan Pavlov who noticed dogs can be fear (stress)-conditioned or reward-conditioned to learn a behavior by associating a cue, a conditioning stimulus like a bell, whistle or the sight of a reward or punishment, with an actual reward or punishment.

To explain and simplify the evolution of the main brain circuits, I will use the four main functions of brains in vertebrates and mammals, what the evolutionary psychologists and neuroscientists humorously call the 4F's, the acronym mnemonics denoting Food, Fight, Flight, F..k (mating and reproductive sex). Our brain evolved to manage the 4 F's through the 4 P's of risk reduction: Prediction, Prevention, Protection and Planning. Through this process, our brain manages our body's physiology by constantly assessing risks and metabolic costs associated with our pursuit of reward (food and sex). As discussed in the last chapter, intelligence in all life forms has evolved for the purpose of metabolic efficiency, but unlike trees and earlier life forms that rely significantly on photosynthesis, symbiosis and slow metabolism, vertebrates, and particularly mammals, have faster metabolisms that require more complex brains. Our brain and body use nervous system signals and hormones, which are usually peptide (protein) or fat (cholesterol) based molecules, as energy signals. Humans have evolved the following five major hormonal systems, three of them self-reinforcing and addictive, to handle all sorts of challenges posed by nature and other humans:

Reward Conditioning System

As early hominins ventured into a new savanna they had to expand beyond their previous comfort or knowledge zone, which was shared by other primates.

So their starting point of brain development could be the *instinctual* motivation to pursue a yellow-colored fruit (a banana) which is a known source of energy and therefore, an essential reward of nature. So for the early hominin the sight of a certain yellow fruit was a conditioned stimulus that would have his brain authorize work and spend energy in the pursuit of the reward, which is the calorie in the fruit. As discussed in chapters 2 and 3, natural evolution is all about homeostatic baselines and metabolic efficiency so the calculus of learning in the hominin brain is about making sure the amount of energy spent, such as work done to climb up a tree, to secure the reward (in this case the banana) is not more than the energy available in the food, otherwise he would waste energy and be metabolically inefficient. So in his new realm, outside his comfort (learned) zone, the early hominin comes across a banana-looking fruit growing on a tree that is different in height from the safe ones he had learned about. What should he do with the novelty? Should he pursue this new goal? Will it be rewarding enough or full of unexpected risks such as falling from the tree? What is the metabolic (caloric) cost of climbing this new tree? This is a new learning opportunity.

What the brain does in such uncharted territories (proximal learning zones) is prediction and simulation. In other words, a bet! Because any error in this prediction can be costly, curiosity and learning often happens in small steps. Now the reward could end up being actually metabolically easier than predicted, for example if the tree was easier to climb, the bananas were faster to reach (temporal aspect of reward) or needed less work than predicted, or if the new bananas were sweeter, hence more calorically rewarding than expected. In all such cases, the learning brain has to encode this positive *"reward prediction error"* signal to use it for adjustment of future motivation, learning and homeostatic baselines and safety boundaries. The reward prediction error, which is at the heart of reinforcement learning models, is the difference between the *actual* reward experienced and our reward *expectation*[6]. It is like the distance to the bull's eye in shooting, archery or dart ranges, used to train our brain for future calibrations.

In mammalian brains, the task of registering reward prediction errors is assigned to neurons in the midbrain called *dopaminergic*, i.e., producing dopamine, which is the neurotransmitter authorizing and managing goal-oriented pursuit and motivation. Because firing of neurons and releasing neurotransmitters such as dopamine is energy consuming and metabolically expensive, dopaminergic neurons in the human brain have evolved to specialize in "value estimation of rewards." Releasing dopamine would give the *Go-ahead* activation to parts of the brain that authorize our body's motion to pursue a rewarding goal.

The dopaminergic pathway in the human brain starts with the ventral tegmental area (VTA) which is part of several feedback loops innervating (meaning projecting neurons into) large portions of our brain such as:

(A) The hippocampus (which translates as "sea horse" in Greek due to its shape) is associated with mood, memory and contextual attentive learning (not shown in the images because it is buried). It helps us to register the context of the authorized reward for future reference or to pull out past contexts for comparison (see below in Nucleus accumbens). This is why it is essential to remove a rehabilitating addict as far away as possible from places and contexts which trigger the dopaminergic reward memories and joys of being high. All that mood-building context is still there in the hippocampus of an addicted person.

(B) Insula: Part of the cerebral cortex that registers sickness and disgust, so if the banana looks or smells rotten, the insula, based on past experience, will try to override the dopaminergic authorization and prevent pursuit of a rotten reward (banana). Insula has evolved to protect us from pathogens, elicited by substances like feces, vomit and putrid food.

(C) Basal ganglia: It is associated with coordinating our movements in a smooth seamless way. After all, movement towards a goal requires coordinated motor functions. This is the reason why dopamine deficiency could result in Parkinson's disease, a brain disorder disrupting our coordinated motor functions. The concept was brilliantly captured in Awakenings, a 1990 movie starring Robert De Niro and Robin Williams, and inspired by true events about an extreme form of Parkinson's disease.

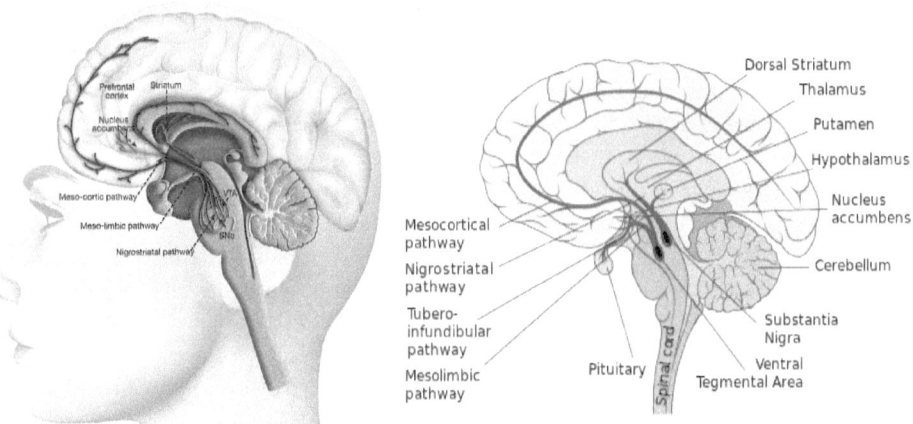

(D) The frontal cortex: This part of the neocortex, which was the last to evolve in *Homo sapiens* brain, is also the last to develop within each one of us, often into our 20's. It is associated with big picture executive analysis and calculation of risk-reward ratios based on inputs received about rewards (dopaminergic) and risks (amygdala discussed in the next section). It often makes the final decision and may override the dopaminergic authorization by projecting back inhibitory neurotransmitters into the nucleus accumbens (see below). An area adjacent to the prefrontal that plays a major role in goal-directed behavior is the anterior cingulate cortex (ACC). New research shows when ACC was switched off in marmoset monkeys, they could no longer make an association between their behavior and a particular outcome (reward or punishment) and resort to perseverative (obsessive, compulsive or addictive) behavior. Interestingly, ACC, which is a bridge between our emotional limbic brain and the cognitive prefrontal cortex, activates when we experience pain or feel other people's pain (empathy). ACC is also believed to be involved in consciousness, self-awareness and free will. So as will be discussed later in the book, consciousness and free will, our ways out of addictions and obsessions, form at the intersection of human cognition and emotions.

(E) Nucleus accumbens: It is part of the basal ganglia involved in triggering our motor functions. It receives authorization signals from the dopaminergic system but also checks it against excitatory or inhibitory confirmation signals from the frontal cortex, insula, amygdala (threat and risk assessment, to be discussed shortly), and hippocampus before triggering our motion.

In our early hominin example, once his dopaminergic system authorizes the pursuit of a new reward, he has to work and spend energy to reach that reward. In nature most rewards were within reach, both temporally and spatially, so metabolic calculations related to motivation were also for the immediate or near future (an example was shared in chapter 2 about a fox chasing a rabbit). Once the hominin receives the visual cues of bananas, the dopaminergic feedback loop connecting the frontal cortex and the hippocampus runs the metabolic calculus comparing the estimated reward value (based on memory and parameters like the size of banana) to the estimated risk and cost (the distance to the banana, and the work needed to climb that tree). If the pursuit is net energy positive and deemed metabolically wise, dopamine is released as the energetic authorization for motor functions and pursuit. As the hominin climbs the tree, the part of his brain called hypothalamus releases glucocorticoid, the hormone accelerating the release of glucose in the body in times of sustained energy need.

Over time, an adaptive brain evolves metabolic efficiency by forming self-adjusting homeostatic baselines and safety boundaries through negative (self-adjusting) feedback loops as discussed in chapter 2. Diseases could be defined as disorders caused by metabolic inefficiency and chronic allostasis (disruption of homeostasis).

The Self-Reinforcing Dopaminergic Brain Circuit

The early hominin's brain evolved the circuitry for curiosity by venturing into new realms and *learning* to survive in new ecosystems and by metabolically assessing the duration, path and outcome value of any reward pursuit. With each new pursuit, our brain would use the *reward-prediction error* to recalibrate its baselines and safety boundaries. The dopaminergic system evolved as an amazing *self-correcting* feedback loop for metabolically efficient pursuit of happiness, initially new caloric or reproductive (mating) rewards. But with time, our frugal brain used the same motivational (dopaminergic) circuits to help us pursue *any* goal oriented activity regardless of its caloric or reproductive value.

Also, over time, the following features made the human brain's dopaminergic reward system prone to forming self-reinforcing addictive feedback loops:

Scale-Free Habituation

Like all other naturally balancing homeostatic or steady state systems discussed in chapter 2, the dopaminergic system has to build and adjust set points and baselines for different rewards. But because the pursuit of a reward is metabolically costly, the brain of humans, as a vagabond species, had to evolve ways to spread out our available and limited motivation and exploration energy among different rewards in proximal learning zones. Otherwise, our early ancestors would use up a lot of energy on the same old pursuits and not be motivated (energized) enough to adaptively learn new concepts or pursue rewards in new ecosystems beyond Africa. This is called *habituation*.

So the motivational learning brain of the early hominids ensured they would not get as excited with the same old banana trees. The brain still motivated them when they were cued with banana sightings but adjusted the reward value baseline downward because the fruit had now become a *familiar* reward as opposed to a *novel* reward for which the brain reserves the valuable supply of dopamine. Remember the firing of dopamine is costly and has to be constantly optimized because it is the foundation of our motivation and movement, i.e., life.

It has indeed been shown[7] that the dopamine's role in cue-reward associations will depend on *baseline* working memory and dopamine synthesis. The downward adjustment of reward baselines ensures the *novelty* wears off for normal rewards. This is why our hominid ancestors had enough energy to explore new rewards and territories and continue learning. The amount of dopamine released would only increase again if the new sighting of bananas signals a *larger* cluster of bananas or a shorter or less distant tree, all of which translate to a more metabolically advantageous ratio of reward (calories gained per unit time) to risk (calories spent per unit time). The brain, based on the old baseline, is telling us *"This level of reward to risk ratio is novel, hence more exciting and worthy of your passion (energy) to pursue than the ordinary small cluster of bananas on tall trees!"* So low hanging fruit becomes our new and exciting purpose, but not for long because again, the dopaminergic system will soon upregulate the "banana on a tree" reward baseline to include and normalize the low-hanging large banana clusters. Our brain forms many of these familiarity baselines, i.e. habituations.

Habituation, which happens in several brain circuits, is adaptive and critical for our survival and continued learning according to the Ecological Intelligence Hypothesis. Without habituation, each time the hominin went to the nearby banana trees he had to engage a lot of brain power to do the whole risk-reward analysis. To use a modern day analogy, without habituation, each time we drove to a local supermarket to buy food, we would have engaged a lot of cortical and limbic brain power to decide on the reward's worthiness and the path. We would even have to relearn how to drive. But thanks to habituation, driving which is coordinating our motor functions is handled by energy efficient canned algorithms that run in our basal ganglia once habituation kicks in and our learning baselines and comfort zones are re-adjusted.

Habituation is adaptive because it works with negative reward prediction errors too. In other words, even if we are not rewarded as expected in life, our baselines are adjusted downwards so we still remain motivated to live. A prisoner looking forward to the day of release or to the next warm meal is doing so because the dopaminergic system has adjusted his reward expectation baselines downward so he is habituated to remain hopeful and motivated to pursue even small rewards. Habituation is one of the reasons we keep our hopes alive even when downtrodden or set back in life. Habituation is what makes sure the brains of the poor still find joys in life like the brains of the rich.

Habituation is also the reason for the *Grass is Greener On The Other Side* feeling and for taking our blessings, like health, friends, family and freedoms, for granted. It is the reason for John Lennon's *Watching the Wheels* song which was released in 1981 posthumously after his assassination. When Lennon decided to step down from his money-making tours in order to slow down and spend more time with his family, his fans and friends were shocked: "*Don't you miss the big time boy, you're no longer on the ball?*" His response as he puts it in the song: *"I'm just sitting here watching the wheels go round and round! No longer riding the merry-go-round"* Neurologically speaking, by "Wheels going round and round" he was referring to addictive cycles of habituation to fame and money that he was stepping away from[vi]. The merry-go-round is an ingenious metaphor for the self-reinforcing and addictive nature of dopaminergic feedback loops.

Habituation can also destroy humans, a main theme of this book. Unlike natural feedback loops that are homeostatic, i.e., self-adjusting, habituation is a form of a positive feedback loop. If you are familiar with addiction, habituation is exactly the reason for the *dependence* and *tolerance* stages of substance addiction. When last year's *high* becomes a new *low* we need more dopamine pumped to stay motivated and enjoy life. That is done by substances like cocaine that act as dopamine agonists, which means they bind to the dopamine transporters on neurons and block it's removal from synapses (spaces connecting neurons). As a

---

[vi] Unfortunately, The Beatles and Lennon had a harder time quitting other addictions (drugs, etc.)

result, because there is more dopamine available in the synapse, an amplified reward signal is sent to the postsynaptic neurons, which causes the temporary euphoria experienced immediately after using cocaine or similar drugs[vii].

There are countless examples of how humans go to extreme lengths and pay high prices for their dependence and addiction to dopaminergic novelty. Last year, a 102-Year-Old Maryland World War II veteran went skydiving despite her fragile posture. In her own words, she was looking for *"The Thrill of a Lifetime."* Neurologically speaking, *thrill* is a code word for seeking dopamine regardless of the risks. Consider Clifton Maloney, a wealthy investment banker, former VP of Goldman Sachs, and husband of U.S. Congresswoman Carolyn Maloney. He seemed to have it all, money and political power. He was known to be an adventurer and proud to be among the oldest Americans who climbed the world's tallest mountain peaks. In 2009, he died while pushing his luck one last time to climb the world's sixth-tallest peak, Cho Oyu in Tibet. Neurologically speaking, he was habituated to his wealth and power so his dopaminergic circuits would prompt him to seek higher rewards in higher risk thrills.

An area of addictive habituation prevalent among financially successful men is sexual addiction which is particularly difficult to quit because it engages two self-reinforcing hormones (dopamine and testosterone). I will discuss this more in the section on testosterone.

So the dopaminergic system that was evolutionary adaptive for the hominin in the new ecosystem (according to Ecological Intelligence Hypothesis) and metabolically optimized for continued learning, i.e., pursuit of novelties made him prone to tolerance and addictive behavior.

Addictive tolerance, in scientific terms, means the dopaminergic systems and reward baselines are logarithmically scale-free (non-linear). In other words, the extra dopamine released seems to correlate with the *ratio* of the sensed reward to predicted reward. So when we expect 5 reward units (say calories in bananas) and receive 10, it is as dopaminergic and motivational as receiving 100 units of reward when we expect 50 units[8]. This is why I felt more of a thrill when my graduate school scholarship went from $500 a month to $1500, than 10 years later when my corporate salary went from $7000 to $10000 a month.

With scale-free habituation, our hominin ancestor was no longer passionate about low-hanging fruits. The novelty had to be a lot more rewarding to shift his logarithmically scaled reward-prediction-error calculations. You can say he was starting to be a bit spoiled. This is how we can spoil our children too, by constantly offering them higher rewards. And habituation is the neuroscience behind both addiction and insatiable appetites (greed). *The Sky's The Limit* is a human-centered motto rooted in our brains' evolution[viii] in response to selection pressures of exploring new habitats.

[vii] There are also pharmaceutical dopamine agonists, called reuptake inhibitors, as well as opioid antagonists like naloxone that trigger the opposite effect, i.e., withdrawal.

[viii] The logarithmic scale-free nature of this should remind us of natural fractal patterns discussed in chapter 3, such as logarithmic spirals.

Delayed Gratification

The brain receives all sorts of dopamine signals with different reward magnitudes and difficulties. Although still not fully understood, our brain's temporal management of reward signals, i.e., how to judge rewards based on the length of time it takes to reach them, seems to be rooted in neurotransmitters such as serotonin that balance dopamine action, and the two major types of dopamine receptors that balance each other:

(A) The D1-like receptors are *excitatory* and not that sensitive but respond rapidly to high dopamine concentrations occurring during *phasic* firing of dopaminergic neurons. This is a burst of immediate and high value reward but of short duration. Many drugs activate this pathway. For example, the euphoric high (rush) from cocaine, crack (powder) cocaine and injected (intravenous) heroin peaks within 15, 5 and 0.1 minutes, respectively. So the D1-like receptors in the nucleus accumbens are the most sensitive to injections which provide the fastest high, followed by an equally spectacular decline[9].

(B) The D2-like receptors are *inhibitory* and respond to low dopamine concentrations present during slow *tonic* firing of dopaminergic neurons. These neurons condition us to enjoy smaller longer-term rewards slowly and to pursue them for longer. Proper homeostasis in a healthy body relies on a healthy balance of D1 and D2 type activations and distribution of motivational energy between phasic and tonic rewards. Studies show that long-term exposure to drugs of abuse such as cocaine which primarily activate D1-like receptors lead to a decrease in tonic D2 receptor signaling and result in drug withdrawal, compulsive intake, and reinstatement of drug-seeking behavior. This same reduction in D2-like receptor signaling is also seen in obesity, internet addiction, and trait impulsivity. Without D2-receptors putting a brake on our urges, we would want all our rewards instantly.

Under normal conditions, a healthy balance of D1-like and D2-like activation exists to condition us to pursue both immediate and distant rewards. Remember life is about balance and we need to pursue both types of rewards. Our advanced powerful frontal cortex, with an ability to conceptualize and imagine worthy long term rewards, can then mediate a gradual release of dopamine over time to help us pursue long-term rewarding goals such as college degrees, vacation homes, million dollar retirement accounts, or bliss and heaven in the hereafter. We will discuss the role of D1 and D2 type receptors in pair-bonding, social relationships and trust later in this chapter.

Interestingly, D1-like and D2-like receptors condition us to be far-sighted or near-sighted in life, literally speaking. In the eye, the dopamine released upon light stimulation regulates homeostatic circadian rhythms (such as retina development, visual signaling, and refractive development). Children that spend more time outside experience delayed onset of myopia (short-sightedness). A working hypothesis of homeostatic control of myopia has been proposed in which activation of D1-like receptors leads to hyperopia (farsightedness) and activation of D2-like receptors leads to myopia[10]. Here again, health is in balance.

A cautionary note: Even humans that choose a delayed gratification reward-seeking path in one area can still be vulnerable to addictions. In fact, many tireless seekers of fame or fortune make up for their delayed gratification and dopamine trickle along their long journey with binging on dopaminergic activities elsewhere in gambling, sex, food, alcohol or drug addictions.

Temporal Discounting, Impatience, and Our Time Perception

Interestingly, because our dopaminergic brain circuits assess the path, duration and outcome of any reward pursuit through time and space, our perception of time is closely related to our anticipation of reward. This is a typical sequence of events inside our body when we are subjected to a reward stimulus:

Usually within 0.1 second, conditioned stimulus and cue is followed by release of dopamine which authorizes the body to pursue the goal[ix]. This is followed, usually within 0.5 second by work and energy spent by the body for the duration of pursuit, which may be followed by achievement of reward and a burst of dopamine depending on the reward prediction-error (surprise factor).

If we did not move our bodies (in space) or pursue rewards, our perception of time, which is linked to our energy metabolism, would be extremely different. So it is reasonable to assume that our brain's perception of time is dictated by our body's *internal* timing and budgeting of its energy including that spent on reaching rewarding goals. The longer it takes to reach rewards, temporally (time) or spatially (distance), the smaller is the assigned reward value by the dopaminergic system. This is called the reward's *subjective value*. Studies show rats, pigeons, monkeys and humans are biased towards smaller and earlier rewards versus larger rewards that are later. Thus, the *subjective values* of rewards shrink when they are delayed even though their *objective value* remains the same.

The decrease of dopamine response with reward delay is indistinguishable from the decrease with lower reward magnitude. Let's go back to our hominid ancestors. If he was habituated with pursuing a reward of 10 bananas on a tree 10 meters high and it took 10 minutes to climb that tree, the wait (reward delay) was 1 minute per banana. Now if he sees a tree that is 15 meters high and hence takes 15 minutes to climb to reach the reward, he would not be happy with a reward of 15 bananas (still 1 per minute of work) because longer pursuit means higher risk. He would probably go for it if the tree offered 20 bananas. This increase over time in the expected reward rate is called *"temporal discounting."*

[ix] The actual mechanism of authorization is often dis-inhibition, which means in the neutral homeostatic state, the central nervous system, including the brain's thalamus, motor cortex and stem, send inhibitory (GABAergic) motor signals to avoid unnecessary body movements. The dopaminergic projections often disinhibit (dampen) these inhibitory signals.

The more the accumbens area is activated and the less the frontal cortex is involved, the steeper a brain's preference for shorter-term pleasures and the less our inclination for delayed gratification. An overactive pleasure-conditioned accumbens overestimates the *subjective value* of reward and an underactive frontal cortex overestimates the *duration* of the delay, both leading to temporal discounting.

Why is temporal discounting important? Because it leads to impatience, which is responsible for both our exponential growth in technological innovations and our addictive behavior and inability to blissfully enjoy the present moment. Temporal discounting makes it more difficult for us to condition our D2-like dopamine receptors with tonic rewards and to curb our addictive habituations.

Temporal discounting is why it is hard for older people to form new habits unless the rewards are large or imminent. Temporal discounting is also why we may recall childhood as a time of bliss when our reward expectation baselines were low. We patiently saw the same rewarding movie several times and enjoyed it each time. We can assume the early hominins pursued goals in a fashion similar to modern day children, by remaining curious and motivated by exploring in small increments.

In financial markets, temporal discounting is the psychological underpinning of Treasury bond yield curves. The government historically offers better annual percent return to investors who buy longer term treasury bonds (like 10 or 20-year bonds) than those who buy bonds that mature earlier, say in a year[x].

Temporal discounting is also the reason behind drug withdrawal symptoms. As addictive habituation progresses, even small delays in the dopaminergic reward (cocaine, sugar, gambling reward, sex, money, etc.) evoke impatience and serious suffering. Perhaps no business has mastered this aspect of the human psyche better than casinos. Modern gambling algorithms that blend low-hanging fruit (small easy wins) with programmed uncertainty in reward timeline and value (stimulating amygdala for emotional drive), and with near-misses are extremely effective in getting people hooked. So the next time you play in a casino, do not get too excited or push your luck when you see two out of three colorful fruits or lucky 7 or dollar signs line up. It is just a ploy to fire up your dopaminergic circuits.

Psychosis

Given the self-reinforcing addictive nature of the dopaminergic system, it should not come as a surprise that patients with schizophrenia (hallucinations and delusional thoughts) show increased dopamine function in the subcortical and basal ganglia regions of the brain. In fact, a large number of antipsychotic medicines have dopamine-receptor antagonistic effects.

---

[x] In rare cases when government focuses on shorter term borrowing and is uncertain of intermediate future outlook (say 1-10 years) we see yield curve inversions

A type of psychosis known as the *Jerusalem Syndrome* involves obsessive delusional experiences following a visit to a holy religious city like Jerusalem. A crisis of euphoric spirituality happens when a religious pilgrim, taking a long journey to bask in the envisioned holy aura of Jerusalem, is gravely disappointed by the mundaneness of Jerusalem, its streets and residents. The shock, basically a devastating reward prediction error in the dopaminergic system, can lead to delusional and psychotic behavior, similar to withdrawal symptoms in addicts.

As humans, we impose a meaning on our world by pursuing abstract conceptual rewards. We face a mental and even physiological crisis when the reward is not a match to the metabolic energy expended. And it is not only religious people who pursue conceptual rewards in life. Political power, fame and even money (in its modern fiat unpegged form) are all conceptual rewards. We will discuss further the concept-driven nature of human brains in chapter 5.

Dopamine is a central link connecting our explorative and learning behavior to our habits, joys and motivations. So it is not a surprise that some of the most driven explorative humans are also the most susceptible to addictions and psychosis. The tabloid industry is mostly built on reporting the never-ending stories of relapses and rehabs among famous folks. Although many stories of addiction and psychosis remain untold, stories like that of Roseanne Barr, shared in chapter 1, are not rare or unique.

Another famous sufferer of psychosis, Howard Stern is a self-confessed sufferer of obsessive–compulsive and other psychotic disorders. In 2019, Stern who is known for his vulgar narcissistic style unabashedly revealed on TV: *"I've gotta tell you, I'm the poster boy for psychotherapy."* Stern admitted[11]: *"My head is on fire. I am completely drained. I used to wash dishes for a living. That was so much more enjoyable, doing physical labor, using your body."* He is confirming our earlier assertion that constant firing of neurons in the dopaminergic system is metabolically expensive and exhausting when we pursue rewards that do not require physical energy.

Stern explains why he continues his vulgar dehumanizing style in his talk shows: *"I couldn't stop myself. It's a horribly neurotic thing….when I don't wear dark glasses and you can see my eyes .. my eyes go back and forth like a mile a minute."* Everything Stern is sharing here fits the description of *perseverative* behavior, robotic thoughtless habituation unpausable by the wise frontal cortex because the dopaminergic system has gone into a dysfunctional self-reinforcing feedback loop. As we shall see later in this chapter, perseverative zombie-like behavior can also be triggered by fear-driven anxiety-induced self-reinforcing loops.

Interestingly, similar to Roseanne, Stern's neurosis and psychosis did not stop him from building his empire and making an estimated $80 million a year, thanks to an expanding fan base. There is a huge market for narcissistic, addictive personality types and obsessive influencers like Stern, who enrage and entertain us. What does that say about us? The more stressed and anxious we are, the more dopamine and novelty we crave. This is what Stern says about why he continues his vulgar dehumanizing style in his talk shows: "*My audience wants something new.*"

But society has not always been rewarding or agnostic towards people suffering from psychosis. In the first seven decades of the 20th century, medical authorities worldwide performed more than 100,000 lobotomies to interrupt dopamine signals in the brains of people they subjectively labeled as psychotic. As we will discuss in Chapter 7, the dangerous procedure led to numerous deaths and disabilities.

Fight or Flight: Threat and Fear-Conditioning System

The brain's dopaminergic system regulates our goal-oriented behavior by calculating, mainly in the frontal cortex, the ratio of reward value to risks/costs. But where does the brain get the data on risks and costs? A major cost associated with the pursuit of reward in the early hunter gatherer hominins was the work (energy) needed to reach the reward, whether it was to climb a fruit tree or hunt game. But there were often other costs and risks associated with hunting or gathering, such as the chance of fighting off predators, the pains and bleeding associated with climbing a thorny tree, the risk of falling in a deep ditch or ravine surrounding the reward.

An essential part of the brain involved in risk assessment is the amygdala, which translates as *almond* in Greek. It is a small, almond-shaped complex of nerve cells that receives, perceives and processes aversive (non-rewarding and costly) fear-provoking or pain-inducing sensory and conceptual stimuli from both the olfactory system and the cerebral cortex. It then associates the pain, anxiety, fear, uncertainty and other risks and costs with a memory, which it stores for Pavlovian type fear conditioning (discussed earlier). The amygdala appears to retain the *episodic* memory and leave the contextual memory of the danger or risk to the hippocampus, which as we learned earlier, also retains contextual settings of the reward.

The amygdala's central nucleus is evolutionarily older so it encodes innate and instinctual fears and dangers, including those we inherit, such as fear of the dark or snakes. The more recently evolved basolateral amygdala, when activated, forms and retrieves conditioned (learned) threat memories.

Despite its small size, the amygdala is one of our brain's busiest exchanges controlling many of our emotions, judgments and body's overall state of relaxation. Its activation directly impacts the dopaminergic and pleasure systems, and vice versa. The physiology of the amygdala involves other brain parts as follows[12]:

(A) Thalamus: It is our brain's sensory relay center. Each time our eyes receive light signals, they send them via the thalamus to the visual cortex of the brain to decode the image. But the amygdala is privileged because it receives sensory signals directly from the thalamus. That means in times of immediate danger, the amygdala acts within a split second without waiting for decoded signals from the sensory cortex or other cortical cognitive deliberations. For example, if the visual input received by the thalamus resembles the image of a venomous snake, motor and sympathetic nervous actions are immediately authorized. The pathway directly activates the central nucleus, amygdala's innate, instinctual evolutionarily inherited risk-assessment core.

(B) Sensory Cortex: The amygdala also receives, with a delay, decoded stimulus input from the sensory cortex, the brain's decoding center for sensory input, for validation of the exact nature of the danger or risk. Given enough time by the amygdala, the visual cortex may decode the snake-looking visual signal as a dead tree branch or a shoelace and hence quell the excitability of the amygdala.

(C) Hippocampus: It stores and communicates to the amygdala, memories, patterns and contexts. So if the fast-action stimulus from the thalamus and the contextual input from the hippocampus confirm imminent risk, the amygdala may not wait for the cortical validation before authorizing fight, flight or freeze action. But if the threat is not deemed as imminent, the amygdala, like many parts of the brain, consults with the deliberative, calculative, and analytical executive in charge: the frontal cortex.

(D) Frontal cortex: This is our cognitive brain. It *thinks* about the fear- or anxiety-inducing stimulus by analyzing potential short and long term risks weighed against all other data such as the dopamine reward signals, before sending a final decision to both the amygdala and dopaminergic system. Because the frontal cortex is deliberative, it is less effective when there is little time for our reaction to the threat. Recent studies have shown the faster we flash images of unfamiliar human faces, especially those of different races or colors, or of people looking stressed, fearful, angry or threatening, the more we activate amygdaloid anxiety and stress signals in the brain. This is why some police officers, faced with escalating situations and short reaction times, perceive mobile phones as guns. In such cases, the threat or risk assessment circuits in the officer's brain will not wait for validation by the sensory or frontal cortex. The more fatigued our brains are, the more time we need to properly assess risk and rewards.

(E) At the same time projections from the prefrontal cortex to the amygdala enable us to exercise conscious control over our anxiety, they also can create anxiety by sending conceptual and *predictive* threat signals. These could be images or thoughts of *predicted* dangers that do not physically or currently exist, such as a demotion at work, losing a job, failing in a task, dying of an infection or disease, etc.

(F) Hypothalamus: This is the brain's main hormonal release authorization center to control the entire body's homeostasis and metabolic energy needs. When the hypothalamus receives a fear, anxiety or threat stress signal from the amygdala, it *authorizes* release of cortisol into the blood for proper metabolism and delivery of energy in the body[xi]. But the release of cortisol is a relatively slow process. For imminent threats, the hypothalamus activates the sympathetic nervous system by sending fast-acting nervous signals through the autonomic central nervous system (our spinal cord) to the adrenal (ad' means above, 'renal means kidney) glands, which pump the quick-action hormone adrenaline (also known as epinephrine) into the bloodstream. Adrenaline, triggers our body's so-called sympathetic *fight or flight* mode which increases the heart rate, dilates pulmonary airways, and constricts blood vessels for faster pumping of oxygenated blood and energy to the body and the brain (for alertness) whether we decide to fight or escape (flight).

(G) In times of fight or flight, *sympathetic* stimulation of alpha-adrenergic (activated by adrenaline) receptors in the urinary tract contributes to urinary continence. Going to the bathroom in the middle of a fight is neither possible nor helpful! Pharmaceutical companies use the adrenergic physiology to encourage urinating. Drugs in the family of alpha blockers, such as Tamsulosin (Flomax brand) help with easier flow of urine in prostate or kidney stone patients by blocking the action of adrenaline receptors and therefore, relaxing the body and urinary tract.

(H) Adrenaline also sharpens our sight, hearing, and other senses. This is how we are able to jump out of the path of an oncoming car even before we think about what we are doing.

(I) Brain Stem: For threats that are deemed serious and imminent, the amygdala also sends rapid distress signals through the brain stem, which controls our body's autonomic (involuntary) functions such as heartbeat and breathing through the central nervous system and spinal cord. An urgent distress signal sent via the brainstem triggers our body's sympathetic *fight-or-flight* mode and has a similar effect to that of adrenaline[xii].

The amygdala, as a core activator of our sympathetic nervous system, makes us alert and stimulated. It is also involved in both aggressive and sexual behavior. There is a deeply rooted cerebral evolutionary link between our fears, anxieties, sexual stimulations and physical aggressions. This does not necessarily mean a correlation between these behaviors. For example, a psychopathic rapist can easily inflict pain on others not because of an overactive amygdala but

---

[xi] Through releasing corticotropin-releasing hormone (CRH) to the pituitary gland which in turn synthesizes and releases adrenocorticotropic hormone (ACTH) which in turn stimulates the production of cortisol by the adrenal glands above the kidneys. Cortisol is a steroid hormone that regulates glucose, protein, and lipid metabolism and suppresses the immune system's response and cellular inflammation.

[xii] Adrenaline (norepinephrine) is the main hormone signaling muscles to contract during sympathetic response. Its precursor, acetylcholine, is our body's key neurotransmitter at neuromuscular junctions converting electrical action potentials from motor neurons into mechanical dilation or contraction of various muscles in the body, hence conversion of electrical to mechanical energy.

because of an underactive one unresponsive to fear and pain. So they are aggressive but not fearful or anxious. On the other hand, one can be fearful and not aggressive. In fact, fear-conditioning is used to domesticate animals, and raise docile human herds.

Circadian Rhythm and Our Stress Baseline and Set point

So let us get back to our thought experiment of imagining an early hominin, likely a hunter gatherer type, in search of food. His brain evaluates the costs and rewards associated with the pursuit using the stress (amygdala) and reward (dopaminergic) systems. On an episodic basis, it is the frontal cortex that usually makes a final decision. However, on a longer term basis, the forecasting brain, to do its 4 P's of energy (and risk) management: Prediction, Prevention, Protection and Planning, needs to predict and budget the body's metabolic needs in advance, as we learned about the dopaminergic systems and habituation.

The brain's risk-based (predictive) energy budgeting process, in many ways similar to our money budgeting process, manages consumption and storage of energy. So the brain has evolved a circadian (daily cyclical) routine and rhythm for the release of cortisol: High in the morning and low at night. This is the reason our blood pressure and heart rate rise sharply in mid-morning. In fact, the hypothalamus which controls the release of cortisol and several other hormones also houses the suprachiasmatic nucleus (SCN), a tiny region sensitive to daylight and responsible for timing and fine-tuning our circadian rhythms and regulating body functions in a 24-hour cycle.

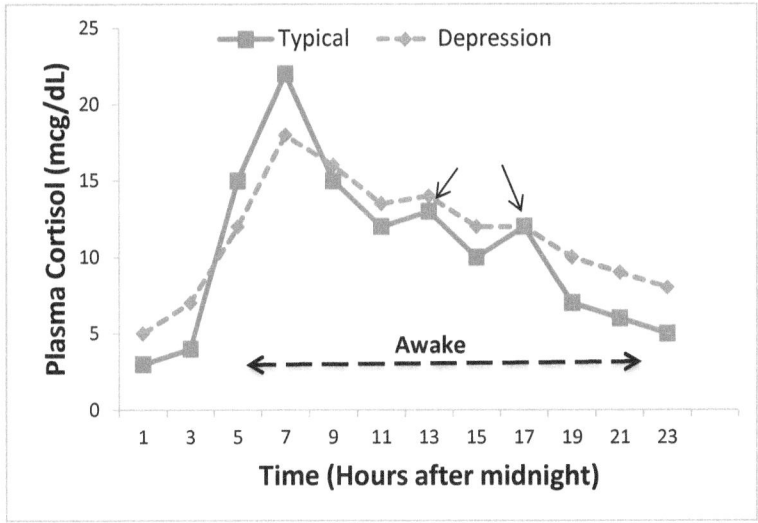

And it is not only the hypothalamus that manages our biological clock. We now know that timekeeping in our body is done collaboratively in all organs by cells talking with each other. It has been shown that the mammalian circadian clock is a coupled network of molecular oscillators within each cell and tissue clock[13]. Light and food (as seen earlier in the dopamine signaling) are predominant cues. It has been demonstrated that dysregulated circadian clocks are associated with metabolic syndrome and cancer. So it is generally wise to listen to our body's naturally evolved cellular-level clocks and not our night owl buddies.

What Does Cortisol Do To Our Body?

Cortisol, as a powerful steroid type hormone, is a primary messenger which means it can cross the cytoplasmic membrane thanks to its fat-soluble properties. Through binding to the glucocorticoid receptors in tissues, cortisol enters the nucleus of the cell and can even directly affect gene transcriptions. As a result, prolonged cortisol release influences all aspects of our body and mind as follows:

(A) Cortisol, like adrenaline, is a *catabolic* hormone which means it switches the body to a "break down" mode and prompts the organs to release stored forms of energy by breaking down fat (adipose tissue) and protein (muscles)[xiii]. Cortisol also acts on the pancreas to decrease insulin - the hormone that mediates cellular metabolism of carbohydrates - and increase glucagon (the ketogenic hormone, discussed in chapter 2) which means the body will reduce consumption of sugar and instead rely more on burning fat[xiv]. This would increase the availability of blood glucose to the brain and body[xv]. So your blood sugar spikes after being stressed, the same way it would spike after eating sugary or high glycemic food. In fact, in many metabolic charts, the impact of cortisol injections and glucose intake are indistinguishable.

(B) As a mediator of our body's sympathetic (pain-sensing)[xvi] nervous reaction, cortisol suppresses major body functions not needed in times of emergency, such as immune response and inflammation. This is why glucocorticoids are medicinally used to treat diseases caused by inflammation and overactive immune systems such as allergies, asthma, autoimmune diseases, sepsis and respiratory viral diseases. Cortisol, as the regulator of stress-induced emergency measures, also suppresses our digestion, cellular growth - by downregulating thyroid function and hormones[xvii] - and reproductive system - by downregulating testosterone or estrogen hormones.

[xiii] Breaking down fat is called lipolysis, and breaking down glycogen is called glycogenolysis.

[xiv] An effect similar to what a ketogenic, high fat - low carb diet induces.

[xv] In fact, intermittent fasting or fasting for 14 hours or more trigger catabolic ketogenic changes in the body that may reduce the risk of diabetes, cancer and heart disease but also upregulate cortisol levels.

[xvi] Greek and Latin roots: 'Sym' means together, 'Pathy' means pains and feelings.

[xvii] The pituitary gland, as the master controller of our hormonal balance, measures circulating levels in feedback loops of ACTH, TSH (thyroid-stimulating hormone), FSH (follicle stimulating hormone) and LH (luteinizing hormone), and adjust their levels in coordination with signals received from the hypothalamus.

(C) The release of cortisol in response to stresses and threats provides our brain with a burst of energy for alertness, so it excites our frontal cortex (judgment), hippocampus (memory engagement) and amygdala (in a self-reinforcing loop, because remember cortisol is often released when amygdala is activated).

What Does Cortisol Overload Do To Our Body?

Similar to the dopaminergic reward-conditioning system, the stress conditioning system, wired through the amygdala, initially evolved as a metabolic balancing mechanism, in this case to authorize the body to spend energy to fight the threat or escape it, all in the *short term*. Our sympathetic response is meant only to temporarily, for a few seconds or minutes, suppress and override our body's major functions such as immune response and inflammation, digestion, cellular growth and reproductive system. For stresses and threats that last beyond seconds and minutes, the cortisol hormone is released into the blood but it also circulates back from the body to the hypothalamus and the pituitary to let our brain know about the body's metabolic state.

This negative feedback loop is self-adjusting which means once the threat is handled and our muscles are done using up the energy (fat and sugar) released by the temporary action of cortisol, the brain reverts back to normal and directs the hypothalamus[xviii] to stop releasing cortisol. This inhibits a *chronic* stress response overdrive in the body. In nature, threats often last for minutes not days so the cortisol feedback loop in the hypothalamus ensures cortisol levels drop to baseline levels when threat (stress) diminishes.

But the human brain has evolved to predict, prevent, protect and plan which means just by *thinking* about future conceptual threats the brain can generate stress levels that are of longer duration, higher frequency and larger magnitude than the natural (4F) stresses handled by the evolutionarily primitive brains in other vertebrates and mammals. *Homo sapiens* brain assigns risk (stress) values to threats envisioned on the path to conceptual rewards such as job promotion, college degrees, beach houses, money, and luxury cars. These are not imaginary but metabolically conceptual because no physical cortisol-dependent *energy* is needed to fist fight or wrestle your way to a job promotion or a nice car or house. Yet, to handle these prolonged *mental* stresses, our brain's main tool is cortisol and the evolutionarily old adrenocorticotropic hormone circuit. As a result, our diurnal (during the day) baseline cortisol levels that naturally evolved to synchronize with our circadian rhythm can be easily thrown off by all sorts of goal-oriented threat-perceptions and stressors. This disrupts coordination among our internal clocks and has been linked to diseases ranging from metabolic syndrome to cancer.

[xviii] The cortisol production signal is sent to the adrenal (above-kidney) glands through the adrenocorticotropic hormone (ACTH) which is released into the blood by the pituitary (in our brain) which in turn is stimulated by the Corticotropin-releasing hormone (CRH) and the cortisol feedback loop.

In the wild, a rabbit, either escapes the fox and then calms down, or becomes food and perishes before he gets into chronic cortisol overload and release. Unlike the rabbit, *our* chronic stress and release of (catabolic) cortisol keeps breaking down (catabolizing) our body.

So what are other consequences of chronic stress and cortisol release? Cortisol, as a powerful steroid type hormone, can enter the nucleus of our cells and affect gene transcriptions. Glucocorticoid receptors are expressed in almost every cell in our body so they regulate genes controlling our cellular metabolism and inflammatory immune response in many ways:

(A) Baseline physiological levels of glucocorticoids help with bone mass accrual but cortisol levels above the baseline cause reduced bone growth and mineral density as seen in Cushing's syndrome and cortisol-induced osteoporosis. Elevated levels of cortisol also reduce intestinal calcium absorption which besides bone density loss can cause neuromuscular irritability, muscle spasms, cramps and twitches. This may explain the role of vitamin D in mitigating effects of cortisol excess. New research also shows that a disruption of calcium metabolism by cortisol will impact our immune system because our adaptive immune system's T cells rely on calcium to metabolize blood glucose for multiplication and clonal expansion[14].

(B) Elevated cortisol levels induce water retention and weight gain in our body through impacting the antidiuretic hormone.

(C) Chronically high levels of cortisol can lead to Cushing's syndrome, also named hypercortisolism, characterized by thin arms and legs, fatty abdomen (pot-belly) and neck (buffalo hump), skin problems such as easy bruising, acne and slow skin healing. Visceral (belly) fat, scientifically (and euphemistically) referred to as the *intra-abdominal adipose depot*, expresses a large number of glucocorticoid receptors so it is more sensitive to cortisol disruptions.

(D) Steroid diabetes and insulin resistance: As already explained, our blood sugar spikes after being stressed, the same way it would after eating a lot of sugar. Furthermore, cortisol inhibits insulin secretion from pancreatic β-cells and impairs insulin-mediated glucose uptake. When chronic or prolonged, this may result in hyperglycemia (high blood sugar) and hyperinsulinemia (too much insulin) and ultimately in insulin resistance. Pre-diabetes and diabetes impact more than 40% of Americans and are now common denominators of comorbidities in several diseases including COVID-19.

(E) Insulin resistance is not the only problem triggered by chronic stress and cortisol levels. Because glucocorticoid receptors are expressed in almost every cell in our body, during prolonged (perceived) stress, our body will decrease tissue sensitivity to the hormone and its anti-inflammatory response in order to avoid the damaging catabolic effects of cortisol. Glucocorticoid receptor resistance has been reported in patients with chronic diseases such as asthma, depression (more on this later), cancer and cardiovascular diseases. Studies have shown[15] that people who suffer from chronic stress may produce more, not less, inflammation-inducing cytokines (chemical messengers) when exposed to viral diseases due to Glucocorticoid receptor resistance. Research by Dr. Sheldon Cohen shows[16] that surprisingly many of the people who were given nasal drops containing respiratory cold or influenza viruses (such as Coronavirus type 229E) did not develop symptomatic colds. Incidence of clinical colds among those infected was associated with the level of chronic physical and psychological (perceived) stress. These populations may be less able to regulate their inflammatory response and the cytokine storm, and therefore, suffer from more severe symptoms when exposed to viruses and pathogens.

(F) The impact of Glucocorticoid receptor resistance and our body's inability to reduce inflammation is exacerbated by the immunosuppression effect of Glucocorticoids, through the decreases in the function and numbers of lymphocytes, including both B cells and T cells.

(G) Chronic cortisol excess also suppresses the reproductive system, by downregulating testosterone and estrogen hormones, leading to disorders such as low libido and erectile dysfunction in men and irregular menstrual cycles in women.

(H) The downregulation of testosterone and the catabolic nature of cortisol would result, in the long term, in reduced muscle mass and repair.

(I) Chronic stress levels and cortisol circulation would negatively impact our digestive tract and lead to chronic constipation or irritable bowel syndrome (IBS), particularly in women.

(J) As a stress mediator, cortisol not only suppresses our digestion but also (in adults and older populations) cellular metabolism and growth across the body so it downregulates our thyroid function and hormones such as TSH, the thyroid-stimulating hormone (as seen in the feedback loop shown). Thyroid is the important endocrine system organ that controls our overall cellular metabolism, and as a result, our heart rate and body temperature baselines. For proper thyroid function, homeostatic feedback loops need to maintain a delicate balance between cortisol, testosterone and estrogen (and progesterone in women) hormones. Because chronic stress and cortisol excess also downregulate testosterone and estrogen, they have a confounding detrimental impact on thyroid function leading to symptoms such as fatigue, cold sensitivity, constipation, dry skin, and unexplained weight gain. The resulting thyroid dysfunction is akin to putting our body in hibernation mode except we not only do not hibernate, but are on an overdrive due to cortisol, glucose and insulin surge. It is the worst of both worlds. It is like hibernating in the middle of a fight with all the adrenaline and sugar circulating in our blood! It is a good example of overloaded feedback loops, which is termed allostatic load[xix].

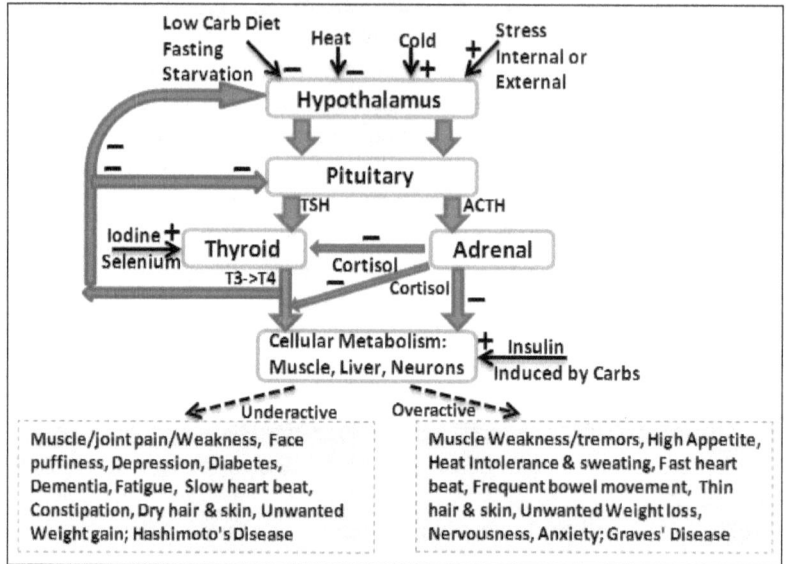

(K) High levels of cortisol can increase blood cholesterol, triglycerides, blood sugar, blood pressure, and the buildup of plaque deposits in the arteries and, therefore, lead to heart disease.

All of the malaise and allostatic load that chronic cortisol overload induces would make it maladaptive for any other species. A fatigued, overweight, reproductively challenged, immunosuppressed wild rabbit with osteoporosis (fragile bones) would not be naturally selected. Yet thanks to our ingenious brains, *Homo sapiens* have found ways, for the time being, to live with chronic stress, cortisol overload and metabolic allostasis despite its very high cost to us and our ecosystem (other species).

What Does Cortisol Overload Do To Our Minds?

The frontal cortex, the brain's seat of cognition, has low affinity receptors for cortisol whereas the hippocampus (seat of memory) and the amygdala (seat of vigilance) have high affinity receptors for cortisol. As a result, behavioral scientists have observed a concave dose-response relationship[xx] between cortisol levels and cognitive sharpness/memory recall performance[17]. Our cognitive performance initially increases with physiological or mental stress (arousal), but only up to a point. When levels of arousal become too high or too

---

[xix] Allostatic load: A term recently coined by scientists to refer to chronic stresses on our homeostatic feedback loops, leading to chronic imbalances or fragile tenuous balanced steady states

[xx] Also called the Yerkes-Dodson curve, named after the psychologists who first discovered the effect

frequent, mental sharpness (memory and vigilance) and physical performance decrease[xxi] owing to the saturation of the hippocampus and amygdala.

The inverted V (or U) shape curve is also aligned with the general concept of hormesis[18], the adaptive (beneficial) response of cells and organisms to a moderate (usually intermittent) stress and their toxic response to prolonged, high frequency or high dose (acute) stimuli.

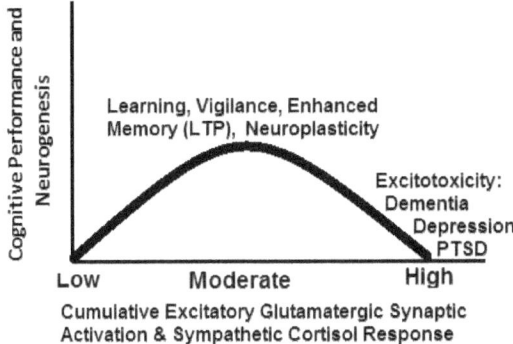

In chapter 2 we discussed how our brain neurons use adaptive strategies such as arborization (tree-like branching) to achieve long-term metabolic efficiency. This amazing cerebral ability to change and adapt as a result of learning and life experience is called neuronal plasticity or *neuroplasticity*. It generally means our experiences in life can expand or shrink certain parts of our brain to meet temporary learning needs while maintaining neuronal homeostasis. Neuroplasticity ensures evolutionary resilience, and both short and long-term metabolic efficiency by balancing energy production (ATP metabolism) and consumption (synaptic activity)[xxii]. Consequently, the energy dynamics over time leads to shrinkage or expansion of dendrites (neurons' receiving tentacles similar to tree roots) and dendritic spines in response to the overloads of stress (energy demand).

Chronic or spiked stress and cortisol levels beyond optimal levels at peak cognitive performance cause disorders such as PTSD (post-traumatic stress disorder) which will expose neurons in the frontal cortex, the hippocampus and the amygdala to prolonged and damaging excitatory (glutamatergic) surges of energy. To protect itself against the dysregulation of glutamate release, the brain resorts to shrinking dendrites in the hippocampus and frontal cortex[xxiii] and

---

[xxi] Interestingly, the effect of dopamine on attention and prefrontal cortical functions also seems to follow an inverted U-shaped curve

[xxii] It is known that neurons, through their synapses, impose energy demands over post-synaptic neurons in a close loop-manner, modulating the dynamics of local circuits

[xxiii] Depression, dementia and shrinkage of neurons are often associated with their over-excitation by neurotransmitter glutamate which in turn is caused by downregulation of presynaptic mGlu2 receptors which help with re-uptake of glutamate and exert an inhibitory tone on glutamate release into the synaptic space.

expanding neurons in the amygdala. These changes have immense consequences for our mental and physical health. The brain's adaptive neuroplasticity, seen in other mammals as well, is a survival mechanism in response to variable environmental demands and selection pressures.

The Self-Reinforcing Fear-Conditioning Feedback Loop

When U.S. President Franklin Roosevelt said *"The Only Thing We Have to Fear Is Fear Itself"* he was making an insightful reference to our vulnerability as humans to paranoia and fear (stress)-conditioning feedback loops. In modern humans, *The Sky's the Limit* does not only apply to our accessible pleasures everywhere and addictive dopamine signaling, but also to our psychological stresses such as job-related or relationships-related stresses. Neurologically, stress is any condition that elicits patterned, compensatory responses in response to the discrepancy between our expectations and perceptions. Both expectations and perceptions can be genetically programmed, established by prior learning, or deduced from circumstances, of the internal or external environment.[19]

And every time cortisol is released, it mediates further vigilance in the amygdala because of its high affinity glucocorticoid receptors. So the amygdala can form a self-reinforcing loop with cortisol under steady or prolonged stress. Chronic stress, PTSD or fear-driven feedback cycles can result in any of the following dysfunctions:

(A) Irrational Zombie-like (*Perseverative*) Behavior: A fatigued, overloaded or underactive frontal cortex and hippocampus, together with an overactive amygdala, pave the way to a state of acting without thinking. In such an imbalanced brain, the frontal cortex, particularly in males[xxiv], cannot project enough analytical *big-picture, long-term* sense, and the hippocampus cannot provide enough context, to talk the anxiety-prone amygdala out of an impulsive action. This is how the stress of wars and conflicts can lead to post-duty PTSD and on-duty impulses. This is why when we are too scared, we cannot think clearly. When the amygdala rules the dialogue with the frontal cortex and hippocampus, we act perseveratively, because we are controlled by the amygdala. We act zombie-like. People may even call us *brain-dead*, which is a crude but somewhat accurate adjective because our frontal cortex activity is downregulated by an overactive, panicky amygdala so our *thinking brain* is practically *dead*.

(B) Perseverative behavior can take two forms. In timid anxious souls with enlarged or overactive amygdala chronic stress leads to *overestimation* of risks and formation of, and rumination in, stronger and longer lasting

---

[xxiv] There are sex hormone differences on stress-induced neural plasticity. In males, stress causes shrinkage of dendrites projecting into the frontal cortex, but in females, interestingly, neurons projecting into the amygdala expand their dendrites in the presence of estrogen. In other words, stress can induce *impulsivity* in men but *vigilance* in women, especially if they are ovulating.

unpleasant memories. Many of the harmful physical effects of cortisol overload are seen in fear-conditioned people suffering from anxiety disorders. Historically, tyrants and manipulators use paranoia and mass hysteria as effective tools to fear-condition large numbers of people into defying their own data-driven frontal cortex (before anxiety paralyzed it).

(C) The second form of perseverative behavior is the opposite of paranoid paralysis, namely reckless sensation-seeking and risk underestimation. This often happens in aggressive, trigger-happy or high-testosterone souls or those with an underactive or damaged amygdala that drives more fight than flight responses. Like in the paranoid folks, the frontal cortex of the reckless folks, normally in charge of *accurate* risk-reward calculations, is also downregulated. That is probably what impacted Tony Hsieh, a young billionaire entrepreneur and author of a best-selling book about happiness who was tragically plagued with alcohol, drug abuse, extreme behavior and a fascination with fire which ultimately took his life. His other extreme behavior included starving himself of oxygen to induce hypoxia and using nitrous oxide, as well as a fasting to the point where he was under 100 pounds. Too much or too little fear, not balanced by healthy cortical regulations, leads to bad decisions.

(D) Although not pathological, the case of a low activity sensation-seeking amygdala may apply to the fearless[xxv] Alex Honnold, whose free solo climb (without ropes) on El Capitan, the 3,200 vertical feet of sheer granite in Yosemite National Park, California, is considered the greatest rock-climbing achievement in history. The route typically takes seasoned climbers four days to complete. That's with ropes. Honnold did it in less than four hours without ropes and by himself - hence the term "free solo," which means rope-free and alone.

(E) A specific case of an underactive amygdala is caused by the process of fear-extinction or anxiety-extinction. This happens when a conditioning stimulus such as the sight of an explosion is repeatedly presented to the subject without the original painful or physically threatening stimulus[xxvi] that was initially or instinctually associated with the fear-conditioning. In such cases a stimulus that triggers a conditioned fear gradually loses its effect. This is likely how a technique called Eye Movement Desensitization and Reprocessing (EMDR) works to treat anxiety disorders and PTSD, and why war videogames and virtual reality simulators are useful to condition armed forces.

[xxv] fMRI scans of Honnold's brain by Dr. Jane Joseph of Medical University of South Carolina showed that his amygdala was almost dormant and non-responsive to usual threatening visual stimuli.
[xxvi] Technically called an "unconditional nociceptive stimulus."

(F) As we shall see shortly, testosterone also plays a major role in amplifying both our impulsive reward seeking drive as well as response by an excited aggressive amygdala.

(G) Memory and Mood Dysfunction: We often have enhanced memories for emotionally arousing events thanks to a brain process called long-term potentiation (LTP; the process of forming stronger longer-term memories), which is optimal when glucocorticoid levels are mildly elevated. In other words, we form stronger emotional memories in times of stress. However, chronic stress and cortisol release, as already discussed, are shown to cause atrophy in the hippocampus, the brain's specialist in memory and mood. This is why elevated levels of glucocorticoids, induced by stress or anti-inflammatory medications to treat asthma have been shown to create deficits in memory and attention, a condition known as "steroid dementia."

(H) In addition to dementia, stress and cortisol[xxvii] may also lead to Alzheimer's disease (AD) by creating an imbalance in the level of toxic amyloid plaques[20].

(I) Depression: Cortisol is a real killjoy for the dopaminergic system, both in the short and long term. The initial burst of cortisol, as the brain's risk signal, generally decreases the activation of the nucleus accumbens in response to rewarding dopaminergic stimuli. This makes sense. Fighting or fleeing consumes energy so stress should make us think twice about energy-consuming dopaminergic ventures. Prolonged stress and elevated cortisol levels are also shown to dysregulate dopamine secretion and the nucleus accumbens, but this does not always lead to anhedonia (suppressed joy and motivation) or depression.

(J) Studies on the mechanistic impact of prolonged stress and cortisol levels on depression seem to be inconclusive. Some studies even found no relationship between elevated cortisol secretion and major depression. The main clue can probably be found again in the brain's metabolic balancing act in optimally budgeting the body's energy through hormonal baselines and set points. If prolonged stress always led to high levels of cortisol and then to depression (lack of motivation), a large number of species would perish in hostile adversarial ecosystems, including humans in war zones and rough neighborhoods, inner cities and ghettos. Research shows cortisol is only one of the risk factors for depression, besides genes, environmental factors and psychological temperament (such as neuroticism and rumination). It is the balance of actions by multiple hormones that determines our psychological health.

[xxvii] Cerebrospinal fluid (CSF) cortisol levels have been found to be significantly higher in Alzheimer's disease (AD) patients than in controls.

(K) For example, in many underprivileged communities, children and adults develop a thicker skin and more resilient temperament in their affectual response to environmental stress, hence they probably demonstrate less stress-induced cortisol sensitivity than residents of higher income neighborhoods. But prolonged adverse childhood events (ACE) or early life stress (ELS) have also been shown to epigenetically lead to stress-induced structural remodeling of the brain. They induce shrinkage of the hippocampus and prefrontal cortex and expansion of the amygdala (in males) as well as glucocorticoid resistance and abnormal cortisol responses that persist into adulthood. The epigenetic change is essential to metabolically protect the organism from cortisol-induced fatigue while keeping him or her vigilant enough to the environmental risks. These changes may lead to trauma-associated psychiatric disorders. Impaired glucocorticoid receptor signaling is a key mechanism in the pathogenesis of depression. The brain, guided by evolution and selfish genes, ensures we survive in war zones and in abusive childhood environments, at the expense of vulnerability to diseases or depression down the road. Nature is always about trade-offs.

(L) As stated earlier, another protective mechanism deployed by our brain to predict and manage the metabolic load of daily stresses in life is a circadian (predictive) rhythm. Any cortisol release, prolonged or temporary, which defies and flattens the homeostatic circadian cortisol rhythms[21] also flattens our motivational energy baselines and throws off our metabolic (circadian) planning. People suffering from major depressive order are shown to exhibit flatter diurnal cortisol rhythms when compared to healthy people. The dysfunction of circadian cortisol levels can be caused by lifestyle factors such as late day eating or exercise, graveyard night shift at work, irregular eating (snacking) in between meals, east-west travel and jet lags, and late sleep and wake up hours that are significantly past sunset and sunrise, respectively.

(M) Abnormal diurnal variation of cortisol is also associated with an upper body fat distribution as measured by waist-hip ratio (a pot belly)[22] and with impaired motor learning skills and glucocorticoid resistance in people with major depressive disorder[23]. Frequent travelers who experience chronic jet lags and disruptions to their circadian rhythms are also shown to suffer from damaging neurological fatigue.

Competition-Conditioning System: Challenge and Social Ranking

Let us get back to our early hominin ancestors. We learned about the reward-conditioning and the fear/risk conditioning neurological systems in the pursuit of food, sex or shelter. But how did our brain condition us for competitors and *others* trying to beat us to food, sexual mates or social status and power? Winning in the social competition does not always involve a fight or flight physical response so our brain needs to use a hormone other than cortisol to condition us for competition and sociality. Besides, cortisol is a catabolic hormone that breaks down our muscles. Evolution needed a muscle-building (anabolic) hormone to balance cortisol and properly condition us for competitive arms wrestling with each other or with other species.

Neuroscientists now agree that the hormonal signal used to stimulate our competitive nature in times of *challenge* is testosterone. In our brain's metabolic calculus, the release of testosterone tips the scale in favor of impulsive bravado type action versus deliberative passive behavior.

In fact, testosterone and cortisol are known to have a mutually opposing effect on our body and social behavior. Testosterone generally favors approach and aggression, while cortisol leads to fear and avoidance. Testosterone is an anabolic steroid hormone so it leads to growth in muscles and bones, but cortisol is a catabolic hormone leading to breaking down of fat and muscles. A healthy homeostasis depends on the proper balance of these hormones so the ratio of cortisol to testosterone in blood tests is a useful biomarker to assess our body's overall catabolic (break down) to anabolic (build or bury) performance[xxviii].

It does make sense that our frugal hormone signaling system uses a sex hormone (testosterone, produced in men's testes and to a smaller extent in women's ovaries) to condition us for competition. After all, much of the primitive competition in nature is related to mating, particularly in tournament species such as gorillas, peacocks and lions, in which members of one sex (usually males) compete in order to mate with the other sex.

Obviously, in modern humans, testosterone stimulation is not entirely for sexual competition and has extended to any other *perceived* socio-cultural or socioeconomic challenges such as job applications and promotions, political races, competitive sports, games or even academics (school grades, university admissions, etc.). In all such cases, testosterone makes us more competitive, and if necessary and pre-conditioned, more aggressive or imposing, particularly towards the underdogs. This is why testosterone brings out the worst in those individuals known to have a "*kick-down, kiss-up*" strategy. Interestingly, if being competitive and attaining a higher social rank means being nice to others, as seen in research on some primate colonies[24], guys with high testosterone levels outcompete each other, not with aggression, but with niceties!

Like many other hormones, testosterone has a context-dependent (contingent) secondary modulation effect on other neural pathways. For example, in young males, spatial ability to analyze relations among objects is highly elevated and is associated with high levels of testosterone. This makes sense in the evolutionary context of winning physical contests in open space. In

---

[xxviii] The difficulty of benchmarking cortisol in blood tests is its diurnal and variable stress-dependent nature

modern life, the skill may translate to solving 3-D puzzles like the Rubik's Cube or mastery of combat or maze-navigation video or virtual reality games.

The Self-Reinforcing Competitive Feedback Loop

Similar to reward-seeking and fear-conditioning, competition, and particularly winning, can be addictive. Testosterone supplies anabolic energy so it is our metabolic hormonal drive to *win* when we are subjected to *physical* challenges. As such, it serves an evolutionary role by boosting our confidence and keeping our fears and anxieties (amygdaloid activity) in check. But unfortunately for humans, even winning could lead to a self-reinforcing addictive feedback loop called the *"winner effect."* Winning in a competition not only triggers testosterone release but it also increases the number of testosterone receptors in the area of the brain through which the amygdala communicates with the rest of the brain. So with each win, the testosterone and competitive confidence building circuit would further suppress the cortisol and fear conditioning circuit. Winning also increases the number of testosterone receptors in the dopamine pathways such as the nucleus accumbens, leading to feeling ecstatic and joyous upon victory. So with each win, we not only overpower our rivals, but we overpower our own reservations and fears. This could lead to overconfidence, ego-centrism, hubris and narcissism.

Testosterone seems to amplify both the reward (dopamine) and stress (cortisol) actions in our brain so it will amplify our vulnerability to addictive pleasure seeking, anxious paranoia or reckless competition. But the cortisol feedback loop, which naturally balances testosterone levels, is shown to get out of whack at the very top and bottom of dominance hierarchies[25]. Among primates like baboons, alpha (top-ranked) and omega (bottom-ranked) males experience the highest level of both cortisol (stress) and testosterone (competition) hormones. The hormonal imbalances in humans resemble that of alpha and omega baboons which should be humiliating for those of us who think we are superior and more civilized than primates. We may not be as brutish as primates but are certainly as aggressive, stressed and restless in our competitions. We just have changed the rules of competition from metabolic (food) and reproductive to economic. Hubris and narcissism are unique human traits. In wild natural settings, rivalries, winnings and testosterone release are all kept in check (particularly by the cortisol feedback loop) because each victory in a physical match involves spending a lot of muscular energy (anabolic testosterone) and often blood and carnage. Most human challenges and victories, on the other hand, are non-physical and certainly not bloody. So *The Sky's the Limit* applies to humans' victories and winning addiction too. Look around and you would see quite a bit of this among top athletes, politicians and businessmen.

An area of addictive habituation prevalent among financially successful men, and particularly difficult to quit because it engages two self-reinforcing hormones (dopamine and testosterone), is sexual addiction. The addiction is the

Achilles heel of otherwise untouchable millionaires and billionaires such as the movie mogul Harvey Weinstein and the late billionaire sex trafficker Jeffrey Epstein. Even the intelligent calculative Bill Gates, despite all his frontal cortex power is not immune from broken testosterone-dopamine feedback loops and addictive behavior which may have contributed to the dissolution of his marriage[xxix].

Why are so many rich, successful people vulnerable to addictive behavior? Because nothing synergizes the release of both testosterone and dopamine like big consecutive wins (money, political power or sex) which will start self-reinforcing neurological cycles of both dopamine and testosterone. Epstein and Weinstein were not alone. The *Me Too* movement by victims of sexual abuse exposed countless prominent names in Hollywood, politics and business. Addictive testosterone-dopamine circuits may be the reason why powerful people engage in activities that are challenging and high risk, even if incriminating.

For humans, winning is now often conceptual and not in a physical sense so dopamine and testosterone are even released whenever we get ahead (even if by cheating) in any race, even when we outperform others in publicity and *win* virtual followers on social media, which can be addictive. Recently, Joselyn Cano, a super-voluptuous Instagram model and influencer who was called by her fans as the Mexican Kim Kardashian, died at the young age of 29. She was known for her abnormal body curves created through surgeries. With each new surgery, she had gained more and more followers, totaling nearly 13 million at the time of her death, caused by complications of her last butt-lift surgery. Cano practically conditioned her own brain with each surgical procedure, in a Pavlovian way, into *addiction to winning* more followers and fame.

The testosterone (winning) loop is our brain's third destructive feedback loop, besides the addictive pleasure-seeking loop and the fear-conditioned anxious paranoid loop. As humans, we are often prisoners to these broken and self-reinforcing feedback loops for most of our lives. We are pushed and pulled between our desires, fears and bravados.

But how about female sex hormones such as estrogen? It appears that estrogen's action on various brain regions is different from testosterone. For example, *prolonged* stress in men which downregulates amygdala-cortex connections and leads to impulsive behavior, in the presence of testosterone's *winner effect* leads to more fights than flights, and to risk-taking and sensation-seeking bravado. In females, however, research shows in times of stress, the sex hormone estrogen has the exact opposite effect of testosterone. It has been shown that under intense or prolonged stress, women's frontal cortex remains in communication with the amygdala and in the presence of estrogen, the amygdala actually becomes more vigilant and risk-averse. So stress plus

---

[xxix] Gates says Epstein was an investor in his vaccine initiatives, but according to the news media, Gates' cozy personal relations with Epstein was a reason Melinda Gates filed for a divorce from Bill Gates.

testosterone means bravado and risk-taking in males, whereas stress plus estrogen means vigilance and conservatism in females.

In addition, studies in women show that testosterone can reduce threat responsivity and amygdala reactivity to anger expressions, and instead potentiate *strategic* social behavior in order to maintain or improve status and reputation[26]. In other words, contrary to its effect on men, testosterone in women increases the motivation to *prevent* a social affront.

Bonding: Sociality and Trust-Conditioning system

So our thought experiment showed how the brain of an early hominid evolved networks and pathways to condition/train him/her for new rewards and explorative learning, risk assessment and threat avoidance, and competitive challenges in social ranking and status. Reward seeking, risk avoidance and competitions are all energy consuming physiological behaviors. But how about bonding and collaborative behavior which are *energy-saving* and therefore, metabolically adaptive and beneficial? As we learned earlier in the chapter, *Homo sapiens* owe their brain size and cortical strength to their eusocial nature, which requires not only competitiveness but also collaboration and communication skills.

In addition to being collaborative task-dividing eusocial species, humans are more of a pair-bonding species than a tournament species. Pair-bonding refers to forming long-term trusting social or sexual bonds between individuals, a rare behavior among primates except gibbons and titi monkeys. Pair-bonding is observed in mammals like prairie voles and in birds such as bald eagles, swans and snowy owls. In tournament species which fight over alpha sexual mating positions, muscular strength is an adaptive trait and infanticide (killing babies), is common. In pair-bonding species, however, infanticide is rare and pair-bonding is an adaptive trait because it leads to metabolic efficiency and energy conservation through conflict avoidance.

According to the parental investment theory by biologist Robert Trivers, monogamy and pair-bonding is rare in mammals because offspring are more attached to the mother through lactation. In bird species, however, where both parents can contribute equally to care and feeding of the young, monogamy is adaptive and seen in up to 90% of bird species. Pair-bonding in mammals seems to be adaptive primarily in harsh environments (food scarcity and predator exposure) that make biparental care more beneficial to the survival of the young.

In humans, unlike hormones which have metabolic roles, neuropeptides (a hybrid of a neurotransmitter and a peptide) seem to mediate our social behavior. The two that facilitate social bonding and trust building are oxytocin and vasopressin, both released by the hypothalamic neurons. Oxytocin, primarily a *prosocial* hormone, is shown to facilitate pair-bonding as well as maternal behavior like uterine contraction during labor, and grooming, nursing and olfactory sensing of offspring after birth. Oxytocin also induces high vasopressin levels in the pair-bonded male, which potentiates *paternal* protective behavior in the father[xxx].

In fact, oxytocin's primary evolutionary role seems to be influencing behavioral responses to social stimuli that are differentiated by their trust, love and bonding salience. For example, oxytocin is the mediator for dopamine release within the nucleus accumbens when a mother rat grooms or licks her offspring or when she is reunited with her offspring after forced separation. Oxytocin also mediates a parent's recognition of his/her own child's face. Oxytocin promotes maternal care because when released during breast-feeding, it is shown to reduce[xxxi] levels of cortisol[27]. Oxytocin has its own stress-reducing effect on the brain that helps promote (metabolic) responsiveness to crying infants even when parents are sleepy or tired.

It has also been shown that oxytocin release is impaired in children with autism spectrum disorder (ASD), a neurodevelopmental condition involving impaired social abilities[xxxii].

When released, oxytocin stimulates receptors in the ventral tegmentum (dopamine-releasing part of the brain), amygdala and hippocampus, all leading to extracellular dopamine increases within the nucleus accumbens[xxxiii], which has

-reinforcing feedback loop very similar to the dopaminergic addictive circuitry but based on the rewards in trust, bonding and intimacy. Let me explain how drug addiction and love addiction are neurobiologically similar but different.

The Self-Reinforcing Bonding and Trust-Building Feedback Loop: Addicted to Love?

There are neurobiological parallels and competition between social attachment/love and substance/behavioral addictions. Oxytocin's impact on pair-bonding is more complex than its impact on maternal behavior. Brain studies in pair-bonding prairie voles show oxytocin induces dopamine release upon mating which acts upon two types of receptors in the nucleus accumbens, discussed earlier in this chapter, one as a mediator of the behavior and another as a reinforcer of monogamy, but the order matters a lot:

[xxx] Oxytocin and vasopressin, like other metabolic hormones, are released into blood circulation by the hypothalamic-pituitary axis. Physically, oxytocin stimulates uterine contraction and milk let down in women. Vasopressin is the antidiuretic hormone, which leads to an increase in blood pressure, vigilance and defensive behaviors such as child, mate and territory guarding, particularly in males.

[xxxi] The mechanism involves lowering reactivity of the hypothalamic pituitary adrenal (HPA) axis, thus reducing levels of stress hormones including adrenocorticotropic hormone (ACTH) and cortisol

[xxxii] There is now some evidence associating autism with childhood infections and/or vaccinations. Young children who have dysfunctional cellular energy metabolism might be more prone to autistic regression between 18-30 months of age if they also have infections or immunizations (vaccination) at the same time, according to a paper by J.S. Poling, et. al, "Developmental Regression and Mitochondrial Dysfunction in a Child With Autism," *J Child Neurol.* 2006 Feb; 21(2): 170–172

[xxxiii] This is shown in rats to induce pleasure effects such as penile erection or yawning and relaxation.

1) Mating-induced oxytocin signals received by the tonic-reward D2-like dopamine receptors (D2R) initiate pair-bonding[28]. However, pair-bonding is inhibited if there is parallel addictive reward conditioning in the phasic-reward D1-like receptors, such as those induced by drugs or other intoxicating habits or attachments to another mate.

2) Once a pair bond is formed (two weeks of cohabitation in prairie vole studies, probably longer in humans), D1-like dopamine receptors are upregulated in the brain that would maintain pair-bonding and get the bonded pair *addicted* to each other.

Research shows a strong overlap between the brain regions and neurochemicals involved in both addiction and social attachment[29]. Under normal conditions, a healthy balance of D1-like and D2-like activation exists to condition us to pursue both high intensity and long duration rewards. Disruption in the balance of D1-like and D2-like pathways in favor of D1-like receptors seems to underlie the behavioral changes seen in addiction and other impulse control disorders. So the order of conditioning seems to matter when it comes to behavior and love reinforcement. If done properly (for example through monogamy and without concurrent attachments), oxytocin released by the tonic (long duration, trusting type) bonding with a mate can have a strong influence in curbing and displacing other addictions.

In fact, exogenous (injected) oxytocin is shown to act on nucleus accumbens to attenuate formation and expression of tolerance to drugs of abuse. Oxytocin also mitigates the withdrawal symptoms of morphine, alcohol and amphetamines[xxxiv] such as Adderall and crystal meth.

Although drug addiction and pair-bonding both trigger self-reinforcing feedback loops with overlapping circuits, there are differences between the two. For example, unlike drugs of abuse, with social bonding every encounter releases oxytocin and vasopressin that interacts with the mesolimbic dopamine pathway to increase the salience of social cues and information. In other words, if we enjoy someone we are socially bonding with, our attention is drawn to sights, sounds, odors, unique behaviors, and other characteristics that identify the specific partner. That is our sensory validation of trust.

The oxytocin-dopamine interaction also means spending a lot of *trusting* or *touchy-feely* time could lead to break up of older pair-bonds and formation of new ones. This may explain an increase in divorce rate when spouses spend more time at work with others than bonding at home. Remember, our brains constantly form new baselines and social bonds, which may lead to habituation.

Oxytocin and vasopressin, whether released through parental behavior or pair-bonding[xxxv], are shown to be specific to social memory[30] and not spatial memory or non-social olfactory memory. These neuropeptides seem to be

---

xxxiv Amphetamines are a group of drugs that sharpen attention, increase energy and confidence, and reduce appetite

xxxv According to a theory vaginocervical and nipple stimulation during human sex acts similar to labor and breastfeeding by releasing oxytocin, which serves to induce bonding between sexual partners.

evolution's neural pathway to socially connect us in a trusting manner to each other. They temper (balance) our competitive (testosterone-driven) and individualistic (dopaminergic) pursuits with the rewards of social bonding, metabolic conservation and trust. And because these neuropeptides interact with the dopamine pathways, they modulate our goal-oriented pursuits and behaviors in life, which means they can help some humans temper their financial or hedonistic ambitions to form families or loving relationships.

The mechanism of social action by vasopressin is different from oxytocin. Vasopressin, which literally means "compressing the blood vessels[xxxvi]" induces

-building and social-bonding, vasopressin says protect your family first. If oxytocin promotes global cooperation, vasopressin says protect your nation first. As usual, we need both of these hormones to remain balanced.

One last note about oxytocin: Because it is released when we hug or visit loved ones or see trusted friendly faces, it activates the neuropsychological benefits of caring hugs and familiar faces in childhood or during times of adversity. A tragic outcome of strict lockdown policies during the COVID pandemic was depriving the elderly and children from oxytocin-inducing familiar faces and social bonds that could have immensely helped them with stress-reduction and healing. It is documented that low plasma (blood) oxytocin concentrations are associated with depression. Oxytocin modulates the impact of corticotropin-releasing hormone (CRH) on the pituitary in times of stress and therefore, downregulates cortisol release and our stress response. That is why hugs, touches and familiar faces, not isolation and fear, can calm us down in times of anxiety and panic.

Let's finish this section about oxytocin and the importance of hugs with a wise quote from Paul Vitti - the mafia chief in the movie *Analyze This*. The fictitious character played by Robert De Niro, shares the result of his own psychoanalysis and childhood trauma with another angry criminal mafia lord: *"You are always angry, maybe because you were not hugged as a kid."*

The Neurotransmitter for Strategy, Patience and Morality: Serotonin

The essential amino acid tryptophan[xxxvii], metabolized by Vitamin D and

---

xxxvi Vasopressin is an antidiuretic hormone, induces a sympathetic nervous response and higher blood pressures, and increases during erectile response.
xxxvii Tryptophan, abundant in foods like whole raw milk, turkey and chicken meat, plays a crucial role in the maintenance of systemic homeostasis in our body because it integrates pathways in nutrient sensing, metabolic stress and immune response. Interestingly, the growth signaling hormone in trees, auxin, also has a molecular structure similar to tryptophan.

and neurotransmitters such as dopamine, testosterone and oxytocin. Some of the important mechanisms of action by serotonin[xxxviii] include[31]:

(A) Serotonin makes us steady and reduces our impulses. Overall, serotonin's role in reward- or competition conditioning seems to be that of *continuity* and behavioral inhibition in an aversive context[xxxix] such as monetary losses or omitted rewards. In simple terms, whenever we encounter defeats or lower-than-expected rewards serotonin helps us develop a *thick skin* and avoid impulsive changes of direction. It is the "Keep Calm and Carry On[xl]" neurotransmitter that balances actions by dopamine and testosterone. Only after repeated disappointments and defeats, serotonin is downregulated by inhibitory GABAergic projections (from the Ventral tegmentum).

(B) Serotonin curbs our enthusiasm and addictions. High serotonin levels seem to dampen and balance the behaviorally reinforcing effects of dopamine. In mice, serotonin mediates impulse control whilst anticipating a future reward. Low levels of serotonin are associated with drug addiction. Paradoxically, drugs that increase both dopamine and serotonin, such as monoamine releasers and their most commonly used variants such as cocaine, amphetamine and Methylenedioxy Methamphetamine (MDMA or ecstasy)[xli] disrupt serotonin's role in facilitating contentment[32]. In such cases, serotonin seems to reinforce *steadiness* in the *addictive rewarding* behavior. Some researchers map serotonin action to *liking* and dopamine to *wanting* a reward.

(C) Neurons projecting from the prefrontal cortex can stimulate serotonergic neurons to influence motivation. So serotonin plays a role in more strategic deliberative pursuits.

(D) Serotonin socializes us. In both primates and humans, serotonin function tends to correlate positively with prosocial behaviors such as grooming, cooperation, and affiliation, and negatively with antisocial behaviors such as aggression and isolation.

(E) Serotonin helps us fight depression. Serotonin deficiency is associated with depression, anxiety and suicidal behavior. That is why the SSRI class of antidepressant drugs artificially boosts serotonin levels[xlii]. Yet SSRIs are linked to sexual dysfunctions and the dampening of function of the dopaminergic reward system   - Americans, perhaps due to a genetic polymorphism in serotonin

---

[xxxviii] Serotonin is released in the Raphe area of the midbrain. Outside the brain, serotonin is found primarily in our digestive tract and helps control our bowel movements and function.

[xxxix] In animals, serotonin may modulate both appetitive and aversive conditioning.

[xl] The motivational motto publicized by the British government in 1939 in preparation for World War II

[xli] Monoamine releasers, after cannabis, are probably the most widely used non-prescription drugs.

[xlii] SSRI: Selective Serotonin Reuptake Inhibitors, increase available synaptic serotonin levels by inhibiting reabsorption (reuptake) of serotonin into neurons

[xliii] Serotonergic agents act similar to noradrenergic (stress-induced) agents in response to erotic stimuli

transporters[33], are less likely to be impacted by major depressive disorders or responsive to antidepressants than Caucasians.

(F) Serotonin makes us more empathetic and moral: Interestingly, brain regions such as anterior cingulate cortex (the empathy center)[xliv] that fire up when we imagine or see cruel acts also have dense serotonergic projections. Serotonin increases the vividness of the harm in these brain regions and plays a reinforcing *moral* role to discourage people from endorsing the harmful action in times of moral dilemma. Serotonin is sometimes called the *morality* or *empathy* hormone, as opposed to dopamine, the *hedonistic* hormone. In fact, serotonin is shown to inhibit both actual harm (as in aggression) and imagined harm (as in moral dilemma).

Our Brain's Chemical Soups: Synergies and Antagonism

Our brain is a busy network of signals carried by chemicals (hormones and neurotransmitters), which amplify or balance each other in a context-dependent way. The ultimate goal is to maintain our body's homeostatic balance and metabolic efficiency. Here I list some interactions between these important signaling molecules that control our behaviors, some synergistically and some antagonistically:

Serotonin-Oxytocin: Both molecules are involved in the control of social affiliative behavior. Social reward is supported by an oxytocin-induced release of serotonin. However, studies show this pathway is impaired in autism spectrum disorders (ASD) patients[34].

Serotonin-Dopamine: We already discussed the push and pull of these hormones in curbing our impulses and addictive habituation or giving up when facing disappointing rewards. Substances that spike both of these, however, make it difficult to kick addictions.

Dopamine-Oxytocin: As already discussed, oxytocin stimulates extracellular dopamine increase within the nucleus accumbens[xlv], which has been demonstrated to reciprocally activate oxytocinergic neurons, leading to more oxytocin release. This is our *feel-good* self-reinforcing feedback loop based on the rewards in trust, bonding and intimacy. Love, parenthood and trusting friendships can displace some of our addictions.

Serotonin-Cortisol: Research shows that stress-induced high cortisol levels consume serotonin (increase its uptake[xlvi]), which leads to depressive disorders. In other words, reducing stress may neurologically and mechanistically parallel the action of SSRI antidepressants, without their adverse side effects.

Oxytocin-Vasopressin-Cortisol: Cortisol levels are downregulated by the actions of oxytocin and vasopressin on the pituitary. As such, the oxytocin

---

[xliv] Also ventromedial prefrontal cortex (vmPFC) and amygdala
[xlv] This is shown in rats to induce pleasure effects such as penile erection or yawning and relaxation.
[xlvi] By inducing an increase in the expression of the gene coding for the serotonin transporter

released after a good hug, being a good parent, or forming a trusting loving relationship can curb many of the harmful impacts of cortisol excess such as diabetes, obesity, depression, dementia and Alzheimer's.

Testosterone-Dopamine-Cortisol Loop, The Vicious Cycle: While social neurotransmitters such as serotonin and oxytocin balance dopamine's reward seeking and cortisol's stress actions, testosterone seems to amplify both the reward (dopamine) and stress (cortisol) actions. So watch out for a testosterone-dopamine-cortisol chemical soup because it often means an overly competitive, stressed, hedonistic narcissistic personality, the kind that stops at nothing to attain high alpha positions in dominance hierarchies. In fact, in chapter 5, I will refer to neuroscience and psychological studies that demonstrate test subjects under the influence of power act as if they have suffered a traumatic brain injury—becoming more impulsive, less risk-aware, and, crucially, less adept at seeing things from other people's point of view. Power is shown to impair a specific neural process, 'mirroring,' that may be a cornerstone of empathy.

Contrary to testosterone, the female sex hormone, estrogen, seems to act as a neuro-protector in times of stress and benefit cognitive tasks such as memory-forming mediated by the hippocampus and frontal lobe.

Dopamine-Cortisol Loop: The impact of cortisol on dopamine release in response to rewards seems to be a U-shape curve. Initially stress seems to decrease dopamine release in response to rewarding stimuli but high stress-induced cortisol levels suppress this relation, and are associated with stronger activation of the nucleus accumbens. Individuals with a high cortisol response to stress might be protected against reductions in reward sensitivity which is linked to anhedonia and depression, but they may ultimately be more vulnerable to increased reward sensitivity, and addictions[35]. So the stress-dopamine combo also seems like a lose-lose situation. Stresses are killjoys for some of us and gateways to addictions for others.

Dopamine-Prolactin: Too much dopamine in nature may discourage motherhood. Another name for dopamine is "prolactin inhibiting factor," and prolactin is the hormone that stimulates mammary glands to produce milk. Lactation is suppressed with too much dopamine, which evolutionarily makes sense because it is not the best time to lactate and breastfeed when one is on the go, driven by goals (dopamine) and focused on pursuit of personal rewards or pleasures.

Endorphins-Dopamine-Cortisol-Oxytocin-Testosterone: Early humans' pursuits often involved stress and pain. To manage these, the brain evolved beta-endorphins[xlvii]. These are analgesic (pain-reducing) neuropeptides, sometimes called endogenous opiates, primarily synthesized and stored in the pituitary. Endorphins released in response to painful stress signals   reduce

---

[xlvii] New research shows our immune cells such as lymphocytes and macrophages contain receptors as well as a releasing capacity for neurotransmitters such as dopamine and endorphin during inflammation.
[xlviii] CRH released by the hypothalamus activates pituitary to release endorphins, which bind to mu-opioid receptors in both the CNS (central nervous system) and PNS (peripheral nervous system)

our sensitivity to pain but also increase levels of dopamine[xlix]. This effect neutralizes some of our pain-induced inhibition because pain and pleasure are the two main balancing forces in our motivations. But when there are very low levels of dopamine, more endorphins are used up to disinhibit (motivate) us, which means less endorphins are available for pain management and we will have a lower threshold for feeling pain. This is why patients with Parkinson's disease tend to have abnormally heightened sensitivity to pain because of their dopamine deficiency. On the other hand, hormones such as oxytocin and testosterone which stimulate dopamine release can help increase our pain tolerance. This is why we see outstanding pain tolerance in competitive athletes (thanks to testosterone-induced dopamine and endorphin enhancement) and mothers caring for their children (thanks to oxytocin-induced dopamine and endorphin enhancement).

The release of endorphins (our *endogenous* opiates) is inhibited by the use of *exogenous* opioids and drugs such as Vicodin, Morphine and Fentanyl. This is why chronic administration of narcotics or pharmaceutical opioids is associated with significant risks of opioid induced hyperalgesia (general sensitivity to pain), drug tolerance and addiction.

Why are Addictions Rare in Wild Nature?

Are other vertebrates as vulnerable as us to the habituating addictive effects of dopamine, testosterone and cortisol? Although research papers link dysfunctional behavior to hormone-induced behaviors in humans and lab rats, addictive self-reinforcing hormone actions in wildlife are not widely documented. This is probably because natural (wild) ecosystems, unlike human-made ones, do not allow binging and the kind of brain conditioning that leads to habitual ruts. There are reasons to believe only humans are cursed with these self-reinforcing habitual cycles:

(A) Addiction is a maladaptive behavior in nature. As discussed in chapter 3, only self-balancing (negative) feedback loops are adaptive in natural evolution. Also, in nature too much dopamine is maladaptive because it leads to psychosis and dopamine resistance, which in turn lead to depression and anhedonia, lack of joy and motivation.

(B) Another reason dopaminergic habituation may be rare in wildlife is the competitive selection pressures limiting the abundance of resources. In other words, even if there was a short tree offering low-hanging fruits and easy calories, it would draw so much attention from birds, bees, bears, and deer that it would not last long. So it would be maladaptive for any tree to be an easy addictive dopamine source for other species.

(C) Dopaminergic addictions are expected to be rare in nature because sugary calorie-dense fruits like figs, very popular with chimps, are

---

[xlix] The mechanism actually involves inhibiting the release of GABA inhibitory neurotransmitters, basically inhibiting the dopamine inhibitors.

scarce and, in natural form, contain fiber and water which dilute the caloric density, namely the dopaminergic intensity[l]. Most natural sources of energy have fibers that use up calories to chew and digest. So the net release of calories (dopamine) from wild natural food is low compared to modern processed calorie-dense food like sugar or high fructose syrup.

(D) Excessive dopamine release seems to be maladaptive in nature. Too much dopamine is known to naturally stimulate severe nausea and vomiting[li]. That is why dopamine antagonists (such as D2 receptor antagonists) are used now as effective medicines to treat nausea[lii]. So in wild nature, the nausea associated with binging and too much dopamine is punitive and memorable enough to be maladaptive. This probably taught early hominids moderation and protected them against habituation to *junk*, easy-calorie food even if they came across it.

(E) The same is true for *winning* and the intoxicating self-reinforcing effects of testosterone. In nature, there are metabolic costs such as high energy needs (metabolic rates) and release of stress hormones (cortisol) associated with fighting, winning and maintaining a high alpha rank in dominance hierarchies. Because cortisol and testosterone both suppress the immune system, alphas often experience immunosuppression and health issues. And alphas in nature do not get medically patched. Alpha male baboons in the wild have short dominance periods and an overall reduced health and longevity compared to Beta and mid-ranking positions.

(F) Too much dopamine in nature may discourage motherhood. As mentioned earlier, dopamine is a "prolactin inhibiting factor," and prolactin is the hormone that stimulates mammary glands to produce milk. Therefore, lactation is suppressed with too much dopamine. As more women become driven by personal goals and careers, there is a large increase in the sale of instant baby formulas that replace breastfeeding. The global infant formula market size was valued at $50.5 billion in 2019 and projected to grow at an annual rate of 10.6%, which is 10-20 times higher than the annual rate of population growth. One downside of this trend that impacts the mothers is foregoing the joy from the *feel-good* oxytocin release induced by breastfeeding, and the

---

[l] Honey is among rare calorie-dense natural treats but it was intended for tiny bees and not evolved for human pleasure or dopamine release! I believe if one could ask our stone age hunter gatherer ancestors for a *wish-come-true* they would mostly yearn for wild honey from a feral bee colony, historically the best natural delicacy known to humans & animals. It is certainly calorie-dense and dopaminergic enough that it motivates humans and bears to risk bee stings and allergic reactions in the pursuit.

[li] As seen in patients with Riley Day syndrome or patients with Parkinson disease

[lii] These antiemetic drugs block dopamine D2 receptors in the brain's chemoreceptor trigger zone (CTZ), an area outside of the blood-brain barrier that receives inputs from blood-borne drugs and toxins such as too much calcium or GI tract poisoning, and as such inhibit the emetic (vomiting) impulses sent to the brain's vomiting center in the medulla oblongata.

oxytocin-induced endogenous dopamine which was shown to protect us from destructive addictive cycles.

(G) In nature, the stress-induced cycles that lead to behavioral and psychological ruts in humans are maladaptive because extreme behaviors such as paranoia or recklessness do not bode well in the wild for a chance of survival.

Chapter Synopsis and References:

We reviewed the neuroscience of how the human brain, evolved to constantly explore, learn and compete, is cursed by being easily conditioned by endless rewards, fears or competition into self-reinforcing cycles of addiction, hedonism, paranoia, anxiety, recklessness, or narcissistic cut-throat greed. As humans, regardless of how free or free-willed we think we are, we often subject ourselves or each other to these Pavlovian conditioning routines. We have denatured into a species that does not require metabolic energy efficiency and balance as an adaptive trait. Our *free will* is often an internal dialogue between constantly-firing self-reinforcing neurological circuits in our brain. Our brain and body are driven by chemical hormonal soups that are strong and addictive. Many of our diseases, habitual cycles, irrational self-delusions, social norms and disorders are simply external reflections of our broken internal feedback loops. In chapter 5, I use a historic timeline to describe how we became a supraphysical self-delusional species.

[1] https://petapixel.com/2021/08/26/umg-seems-to-think-it-copyrighted-the-moon/
[2] P. Amodio et. al, "Grow Smart and Die Young: Why Did Cephalopods Evolve Intelligence?", *Trends in Ecology and Evolution*, VOLUME 34, ISSUE 1, P45-56, JANUARY 01, 2019
[3] M.A. Nowak, C.E. Tarnita and E.O. Wilson, "The evolution of eusociality," *Nature*, August 2010
[4] The Story of the Human Body: Evolution, Health, and Disease, by Daniel Lieberman, Vintage, 2013
[5] Ackerman, Sandra. "Discovering the Brain." *National Center for Biotechnology Information*, U.S. National Library of Medicine, 1 Jan. 1992.
[6] Glimcher, P. W. "Understanding Dopamine and Reinforcement Learning: The Dopamine Reward Prediction Error Hypothesis." *Proceedings of the National Academy of Sciences*, vol. 108, no. 42, 2011, pp. 15647-54.
[7] Soutschek, Alexander, et al. "Activation of D1 Receptors Affects Human Reactivity and Flexibility to Valued Cues." *Neuropsychopharmacology*, vol. 45, no. 5, 2020, pp. 780–785.
[8] Schultz, Wolfram. "Dopamine Signals for Reward Value and Risk: Basic and Recent Data." *Behavioral and Brain Functions*, vol. 6, no. 1, 2010, p. 24.
[9] National Institute on Drug Abuse. "National Institute on Drug Abuse (NIDA)." *National Institute on Drug Abuse*.
[10] Zhou, Xiangtian, et al. "Dopamine Signaling and Myopia Development: What Are the Key Challenges." *Progress in Retinal and Eye Research*, vol. 61, 2017, pp. 60–71.
[11] Strauss, Neil. "Howard Stern's Long Struggle and Neurotic Triumph." *Rolling Stone*, March 2011.
[12] Lovallo, W. R., and T. L. Thomas. "Stress Hormones in Psychophysiological Research: Emotional, Behavioral, and Cognitive Implications (Chapter 21) - Handbook of Psychophysiology." 2000. *Cambridge Core*, Cambridge University Press.
[13] Koronowski, Kevin B., and Paolo Sassone-Corsi. "Communicating Clocks Shape Circadian Homeostasis." *Science*, 12 Feb. 2021.
[14] Vaeth, Martin, et al. "Store-Operated ca2+ Entry Controls Clonal Expansion of T Cells through Metabolic Reprogramming." *Immunity*, vol. 47, no. 4, 2017.
[15] Bailey, Michael, et al. "The Hypothalamic-Pituitary-Adrenal Axis and Viral Infection." *Viral Immunology*, vol. 16, no. 2, 2003, pp. 141–157.
[16] Cohen, Sheldon. "Psychosocial Vulnerabilities to Upper Respiratory Infectious Illness: Implications for Susceptibility to Coronavirus Disease 2019 (Covid-19)." *Perspectives on Psychological Science*, vol. 16, no. 1, 2020, pp. 161–174.

[17] Sapolsky, Robert M. "Stress and the Brain: Individual Variability and the Inverted-U." *Nature Neuroscience*, vol. 18, no. 10, 2015, pp. 1344–1346.

[18] Mattson, Mark P. "Hormesis Defined." *Ageing Research Reviews*, vol. 7, no. 1, 2008, pp. 1–7.

[19] Goldstein, David S. "Adrenal responses to stress." *Cellular and molecular neurobiology*, vol. 30,8 (2010): pp. 1433-40.

[20] Dong, Hongxin, and John G. Csernansky. "Effects of Stress and Stress Hormones on Amyloid-β Protein and Plaque Deposition." *Journal of Alzheimer's Disease*, vol. 18, no. 2, 2009, pp. 459–469.

[21] Herbert, J. "Cortisol and Depression: Three Questions for Psychiatry." *Psychological Medicine*, vol. 43, no. 3, 2012, pp. 449–469.

[22] Lee, Mi-Jeong, et al. "Deconstructing the Roles of Glucocorticoids in Adipose Tissue Biology and the Development of Central Obesity." *Biochimica Et Biophysica Acta (BBA) - Molecular Basis of Disease*, vol. 1842, no. 3, 2014, pp. 473–481.

[23] Jarcho, Michael R., et al. "Dysregulated Diurnal Cortisol Pattern Is Associated with Glucocorticoid Resistance in Women with Major Depressive Disorder." *Biological Psychology*, vol. 93, no. 1, 2013, pp. 150–158.

[24] Sapolsky, Robert M. Behave, *The Biology of Humans at Our Best and Worst*. Penguin Books. 2017.

[25] Gesquiere, Laurence R., et al. "Life at the Top: Rank and Stress in Wild Male Baboons." *Science*, vol. 333, no. 6040, 2011, pp. 357–360.

[26] Buades-Rotger, Macià, et al. "Endogenous Testosterone Is Associated with Lower Amygdala Reactivity to Angry Faces and Reduced Aggressive Behavior in Healthy Young Women." *Scientific Reports*, vol. 6, no. 1, 2016.

[27] Love, Tiffany M. "Oxytocin, Motivation and the Role of Dopamine." *Pharmacology Biochemistry and Behavior*, vol. 119, 2014, pp. 49–60.

[28] Edwards, Scott, and David W Self. "Monogamy: Dopamine Ties the Knot." *Nature Neuroscience*, vol. 9, no. 1, 2006, pp. 7–8.

[29] Burkett, James P., and Larry J. Young. "The Behavioral, Anatomical and Pharmacological Parallels between Social Attachment, Love and Addiction." *Psychopharmacology*, vol. 224, no. 1, 2012, pp. 1–26.

[30] Ferguson, Jennifer N., et al. "Social Amnesia in Mice Lacking the Oxytocin Gene." *Nature Genetics*, vol. 25, no. 3, 2000, pp. 284–288.

[31] Fischer, Adrian G., and Markus Ullsperger. "An Update on the Role of Serotonin and Its Interplay with Dopamine for Reward." *Frontiers in Human Neuroscience*, vol. 11, 2017.

[32] Miyazaki, Kayoko W., et al. "Optogenetic Activation of Dorsal Raphe Serotonin Neurons Enhances Patience for Future Rewards." *Current Biology*, vol. 24, no. 17, 2014, pp. 2033–2040.

[33] Mellman, Thomas A., et al. "Serotonin Polymorphisms and Posttraumatic Stress Disorder in a Trauma Exposed African American Population." *Depression and Anxiety*, vol. 26, no. 11, 2009, pp. 993–997.

[34] Lefevre, Arthur, et al. "Oxytocin Fails to Recruit Serotonergic Neurotransmission in the Autistic Brain." *Cerebral Cortex*, vol. 28, no. 12, 2017, pp. 4169–4178.

[35] Oei, Nicole Y.L., et al. "Acute Stress-Induced Cortisol Elevations Mediate Reward System Activity during Subconscious Processing of Sexual Stimuli." *Psychoneuroendocrinology*, vol. 39, 2014, pp. 111–120.

Chapter 5: A Brief History of Humans and Their Brains: The Fruit that Ushered in the Self-Delusional *Homo economicus* Brain

Epigraph: I study in parallel the timeline of our socioeconomic evolution (history) with our brain's evolution (neuroscience) to illustrate the gradual transition of human brains from a metabolic to an economic mode that ignores natural feedback. I will share a brief history of humankind and interesting factoids such as how one fruit changed the path of human evolution and helped *Homo economicus* subjugate *Homo sapiens*! Because our modern concept-driven brain often dwells outside current physical space and time, we are motivated or fear-conditioned by perceived (non-metabolic) stimuli and are never too far from self-delusions, addictions and irrational prejudices. I share shocking results of interesting psychological studies that demonstrate how self-deceptive our brains can be.

Humans Develop Tools to Ignore Nature's Feedback

In chapters 2 and 3, I shared how natural conservation laws, real-time feedback loops and evolutionary principles ensured biodiversity, metabolic energy efficiency and balance at all levels in nature. I also explained that for humans, balance is no longer an adaptive trait. Humans can ignore natural feedback loops such as pain and injury which naturally disable other species until they heal or perish. Pain management drugs and clinics are now among the most lucrative businesses in America.

The drive to ignore injury-induced pain is particularly salient among competitive athletes such as former bodybuilding champion Ronnie Coleman. His title came in a package that included 13 surgeries (including hip replacement), 14 screws bolting his bones, and two cages holding his back together. He says he wouldn't have it any other way. There are numerous other examples of top athletes (such as Jose Canseco, Andre Agassi, Alex Rodriguez, and Lance Armstrong) who used various injections or medicines to push through their body's natural resistance limits.

Our tools allow us to ignore not only natural feedback but also real-time human-to-human (personal) feedback such as analyses and critiques. We now like to exchange *impersonal virtual* emoji-type reactions, sometimes even among the members of the same household because virtual feedback is easy to ignore.

Because we can afford to be imbalanced, we can live on the edge or have extreme traits which are maladaptive in nature. Many of us are one bill or paycheck away from hunger or one pill, hospital visit or surgery away from death or serious illness. Our colony (society) itself is one crisis or pandemic away from pandemonium.

We now live in the age of *Homo economicus*, which means humans who rely on scaling up production and spending as the main adaptive solution to fix all problems. Even most approaches proposed to tackle the issue of human-caused climate change involve *spending* more resources to curb the damage caused by overconsumption in the first place. Every year in Italy, a large group of people spends 12 weeks to install, sew and later remove some 100,000 square meters of a sun-reflecting tarp over one of the many glaciers in the Alps to slow down its melting. In my estimate, the project costs $0.5-1 million in materials and labor or about $5-$10 per square meter of glacier. Imagine how much it would cost to save all the melting glaciers on the planet with this approach. Yet as humans, we believe we can solve all problems with more spending, more technical tools and more production.

As humans, we can now even biohack our own bodies to bypass nature's checks and balances. A prominent advocate for anti-aging biohacking is Ray Kurzweil, the technologist venture capitalist and a proponent of the *transhumanist* movement who uses possibly every credible technology to remain young. Thanks to biohacking, he looks different each time you see him on TV. He certainly has enough tools (resources, money, and technical gadgets) to experiment on himself and others.

Other technologists are now pushing the use of synthetic gene-based methods to teach our naturally-evolved immune system to *behave* and to defeat all sorts of disease and microscopic organisms, once and for all. This is how a biologist[1] who is a faithful believer in new technologies recently shared his exuberance: *"One problem we will probably face if this RNA dream comes true is a rise in allergies, asthmas and autoimmune problems... But then messenger vaccines may also be a way to tackle these, and teach the immune system to behave. What if 2020 went down in history as the year synthetic biology dealt a mortal blow to future viruses and illnesses in general, rather than the year a virus ruined our health, wellbeing and livelihoods?"* I used to share such enthusiasm but am more humble now after years of R&D (research and development) experience and realizing that we have coevolved symbiotically with microorganisms for at least a million years. Unfortunately, many of today's disorders and imbalances were triggered by our irrational exuberance in yesterday's new groundbreaking technologies (such as antibiotics, pesticides, radiations, plastics and chemicals) designed to *deal a mortal blow* to other organisms.

A Parallel History of Humans and Their Brains

When exactly did humans develop the confidence to teach nature how to behave? And when did we part our evolutionary path from other species? I will draw a chronological parallel between the neuroscience of brain evolution in humans with the history of how we separated our evolutionary path from other species. I occasionally reference peer reviewed studies but my conclusions are mostly my own and heuristic. The premise is that as we developed tools to change our ecosystem, our ecosystems also changed us and the wiring of our brains.

50,000 Years Ago: Eusocial Traits, Tools, Languages and Specialization

As explained earlier, the specific anatomy of our early ancestors, particularly their fingers and relatively large brain (neocortex) size, enabled them to develop tools to hunt and protect themselves. The advent of tools drove eusocial humans to *specialize* in tasks and roles. This was more advanced and diversified than the specialization seen in eusocial bee colonies. Human mothers specialized in nurturing babies for years. Grandmothers specialized in remembering experiences (history and wisdom) for continuity of early colonies. Strong young men specialized in hunting, and agile young women in gathering. Tribe members with better eyesight specialized in being watchmen. Specialization works when members of the colony communicate with each other. We know this happens in bee colonies through the waggle 8-shaped dance. Early humans must have had enough communication skills to ensure the watchmen (with good eyesight), the mothers, the hunters, the gatherers and the toolmakers stayed coordinated for the benefit of the colony.

The most important change at this stage of *Homo sapiens'* evolution was a displacement of adaptive *individual* physical traits and fitness in favor of specialization, tools and communicative teamwork. You could be on the lower end of the eyesight distribution curve but survive among humans if you were a good toolmaker. Your teamwork with a watchman and a hunter could help you survive and pass on your genes. As we saw in Chapter 3, this is not common in Darwinian natural selection where *individual* physical traits and balance are adaptive.

Over time, tools became more complex as toolmakers became more skilled, and specialized hunters became better as they focused more on their hunting skills than anything else. As with bees, early humans evolved amazing survival and adaptation skills at the *collective* level but humans went much further than other eusocial species with specialization and toolmaking at the expense of their *individual* physical traits.

As communication and collaboration with other members of the colony became a matter of life or death for humans, they drastically improved their communication skills, language complexity, and brain size. As explained in chapter 3, evolutionary psychologist Robin Dunbar found a correlation between the size of the brain's neocortex and the size of an optimal social network in nature, called the Dunbar's number. This is why the neocortex is sometimes called the *social cortex* and why early human communities had around 150 members.

But to convey messages, languages required definition of concepts so human brains evolved strong conceptualization and abstraction abilities. When the watchman of a tribe came back to warn others about a fast encroaching pack of eight wolves, he had to convey to others *abstract* concepts such as speed and numbers and *tangible* concepts such as wolves. Over time, colonies with more advanced language and conceptualization skills prevailed. The growth of languages, specialization, social skills and the brain size came at the expense of physical traits such as strong muscles, fast legs, good eyesight, and metabolic energy efficiency.

Regardless of all these changes, the currency of life remained to be energy as in wild nature. The brain of hunter-gatherers was mostly a metabolic regulator that *budgeted* their body's caloric needs, and conditioned them for reward, risk, and competition through metabolic biochemicals such as dopamine, cortisol and testosterone.

12,000 Years Ago: Farming, Surplus, Quantity, Immobility

It is widely believed that around 12000 years ago, a large part of the planet emerged from a long 100,000-year glacial period (the last ice age). Large ice masses melted and gave way to arable fertile land. Probably helped by a combination of luck and curious brains, humans learned they could cultivate and grow certain seeds in arable land to feed themselves instead of relying on the arduous, metabolically exhausting process of foraging that made them perennial on-the-go nomads. As humans learned farming and a group of them specialized in the art, they gradually shifted from a nomadic hunter gatherer life to settling near cultivated land. Humans also learned to domesticate docile wild ox roaming in nearby pastures, for farming and later for meat and dairy production. So the agricultural revolution brought with it a more settled, risk-averse, land-bounded lifestyle while still reinforcing specialization, cooperative language, toolmaking, and a social brain subject to a Dunbar's number of 150.

A more steady and reliable source of food, and less time spent on foraging, allowed farmer women to have more children than hunter gatherers. Despite high infant mortality rates, settled mothers had on average 4.4 children surviving to age 16 as compared to 3.8 children by mobile (hunter gatherer) women, reflecting a relative increase of 15.8% in reproductive fitness of farmers. Farming led to higher population densities, sedentarization, increased contact with neighboring populations, the presence of rodents attracted by food stores, the domestication of animals, and fecal pollution. All those factors facilitated the spread of virulent bacterial and viral pathogens as well as soil-borne helminths (roundworm, hookworm, and whipworm). So as the family and colony sizes grew in the Neolithic (agricultural) age, so did mortality and morbidity (disease) rates associated with sedentarization, excess food and metabolic inefficiency. Research reveals[2] that early farming communities, as compared to their hunter gatherer ancestors, suffered from increased prevalence of tuberculosis, syphilis, the plague, overall immunological stress, and deterioration in oral health.

We can say farming humans traded off quality of life (health) for convenience and quantity (food, family size and body fat storage) of life. But none of these changes caused as profound an impact on human evolution as the rewiring of their brain.

With the advent of agriculture, the human brain had to abruptly adjust its caloric-based goal-oriented risk-reward calculus. Metabolically, climbing a tree and hunting are grossly different from planting and harvesting seeds, herding cattle or walking to the grain storage silo. The brain, as the master regulator of our body's energy budget, was seriously thrown off with the extra easily accessible stored grain (energy) and the surplus of bread or milk.

Surpluses are rare and not easy to carry in the wild. But we know that our brain's dopaminergic system gets hooked on surplus (easy) calories and forms scale-free habituations. So in addition to abundance and quantity, the farming era threw off the conditioning switches in the human brain which led to diseases of excess, addiction, convenience and immobility.

As humans grew their food surpluses and communities, they established towns with protected borders, organized governments and armies, written languages, religions and philosophies. The currency of human life and survival was still energy (calories) but farming required less physical prowess than foraging and more tools, land and numerical skills (to plan surpluses). Our brains and bodies had to rewire for all of this.

5000 Years Ago: Written Language, Accounting and Money, Gilgamesh, Religions, City States, Civilization, Tools and Rules

What would a human do with some surplus grains, a little spare time, communication and conceptual skills? Start trading and extend social connections. This is how the world's first recorded organized urbanization took place in city-states such as the Sumerian city of Ur (believed to be site of the Great Flood of the Book of Genesis and Epic of Gilgamesh, possibly related to the story of Noah's Ark), ancient Persian cities of Susa and Ecbatana, Mesopotamian city of Babylon, and Egyptian cities of Thebes and Memphis. Urban centers were formed by clusters of humans who specialized in farming, fighting (soldiers), tool making, rule making (laws and government), wisdom and conceptual skills (medicine men, holy men and philosophers).

An example of early rulemaking, The Code of Hammurabi was written about 3800 years ago by Babylonian king Hammurabi and included harsh penalties such as death for a variety of transgressions like falsified testimonies and accusations without proof, and adultery. Punishments were particularly harsh if victims were free men, statesmen and priests.

Trading and dealing with larger groups of humans required bookkeeping so arithmetic was invented for actuarial purposes, as well as early written symbols which evolved later into written languages to record other (non-bookkeeping) concepts. Exchange tokens were invented as mediums to facilitate barter (early precursors to money). The human brains had to do a lot more counting and bookkeeping than in the pre-agricultural age.

Homo sapiens were fast outpacing other primates and earlier hominins in reliance on collective intelligence, in levels unrivaled by other species' swarm or herd (flock) intelligence. They were forming *superminds*. In their paper titled *Sharing the World with Digital Minds*, Carl Shulman and Nick Bostrom argue[3] that in eusocial species giant superminds are formed when the aggregate welfare of a colony grows faster than the cost of growing brain power of individual organisms.

These developments had two game changing consequences for the evolution of *Homo sapiens* and our brains: 1) Collective and cerebral traits gradually became more adaptive than individual physical prowess and metabolic efficiency; 2) Land, tools and their ownership were now every bit as important in survival as physical traits or skills such as hunting, foraging or even farming.

In my opinion, no other ancient document captures the adaptive survival challenges faced by our ancestors as does the story of Gilgamesh, an epic poem about a king of the Sumerian city-state of Uruk, who is said to have ruled approximately 4500 years ago. Gilgamesh, who was described as two thirds divine and one third mortal, sought immortality by first defeating a wild-animal-turned-human (Enkidu, which may represent domesticated cows), partnering with him to subordinate the divine bull (representing wild animals) and then proceeding to subordinate the guardian god of trees and cedar forests (Humbaba). This epic conquest of animals and trees by human gods symbolically denotes the start of an evolutionary era in which humans felt a god-like confidence to subjugate other species instead of evolving symbiotically with them.

3000-1000 Years Ago: Large Nation States, Economy, Religions, Philosophies, Money, Caste Systems

In early city states, clergymen, philosophers, bookkeepers, rule makers and landowners often outranked farmers and peasants in dominance hierarchies. In Rome, for example, the lower class in the caste system, were the plebeians who were mostly farmers. The upper ruling class consisted of the patricians, who were land owners, often politically savvy, cunning and good orators (such as Rome's Cicero), army generals and censors who were the bookkeeping equivalent of today's bankers, treasury departments and central bankers.

Physical traits were becoming even less important except in armies. For example, a physically unfit patrician who had good communication, networking and cerebral skills often ruled over a muscular peasant. That did not happen in nature or among early *Homo sapiens*. In the early Persian and Indian civilizations, the dominance hierarchy also consisted, from top to bottom, of scholars and priests (men with strong calculative, conceptual and communication skills), statesmen and warriors, bookkeepers and merchants, and commoners/peasants.

These hierarchies, over time, led to what I call the 3 G's of power trinity in post-agricultural civilizations: Gold (assets, land, and capital), Guns (tools, technology, science) and God (laws, secular or religious rules). Human brains now had to learn how to best adapt to gain dominance in these new adaptive paths. It was no longer enough to be a fast hunter or a muscular farmer. Your *brains* had to be agile to outsmart other humans. In the new era of brain games, humans became the evolution's main selection pressure for each other.

By about 2500 years ago, humans cultivated more lands and expanded their small city-states into nation states and empires - centered in areas currently named Persia, Greece, Italy, China, India and Egypt. The world's first widespread religions, philosophies, languages and sciences evolved. Land and gold were precious assets so major land expansion wars ensued, such as those between Persians and Greeks.

An example of the adaptive nature of toolmaking was the superiority of civilizations that were the first to transition from the Bronze Age to the Iron Age. Around 3200 years ago, the use of iron in ancient civilizations of Mesopotamia, Iran and India set them apart from less advanced or barbaric settlements in Europe or Israelites in Canaan.

The first currency and trade instruments were made out of metals and used across empires. As a result of these developments, humans gained yet more speed, quantity, reach and convenience.

Economic fitness became every bit as important as individual metabolic fitness. So the human brain which had evolved around budgeting the body's energy needs transitioned its calculus to economic bookkeeping. The word *economy* is rooted in Greek words *Oikos* (meaning house) and *Nomos* (bookkeeping). This is what I call the early transition of *Homo sapiens* to *Homo economicus*, the economic man.

As national economies and regional trade grew so did the humans' needs for better communication (languages), bookkeeping (math), specializations (tool and rule making) and social and conceptual frameworks (religions, philosophies and early sciences). The world's first major religions and philosophies such as Buddhism, Confucianism, Judaism and Zoroastrianism began during this period and helped large groups of humans share common metaphysical purpose-driven identities. These religions and philosophies were early antidotes to the psychological and social imbalances that resulted from excesses and surpluses of the agricultural age.

Human-invented socioeconomic hierarchies, as opposed to the natural ones (trophic energy-based and force-based) often became steep and suppressive. *Injustice* became a human-invented concept to describe imbalances and transgressions. People resorted to priests and philosophers for moral solutions to injustices and transgressions. Occasionally, people rebelled. In fact, the world's first (somewhat democratic) republic was formed around 509 BC when citizens rebelled and overthrew the last Roman monarchy as a reaction to sexual transgressions (injustice) by Sextus, the promiscuous son of King Lucius Tarquinius, who forcefully raped Lucretia, a noblewoman who committed suicide in disgrace. Ironically, one of the world's earliest recorded *sexual* transgressions was committed by a person named *Sextus*, without whom we would not have the first republic form of government.

1000-300 Years Ago: 3G Consolidation, Religious Wars

As humans relied more on land, tools and rules, power became more concentrated at the top of the dominance hierarchy with land and asset owners, toolmakers, and secular or religious leaders. But along with the power concentration, came imbalances and conflict. In some civilizations, the lawmakers and priests colluded against landowners. In other civilizations, landowners colluded with priests and lawmakers against the plebes.

The medieval period witnessed consolidation of religious authority in Europe and inter-religion crusade wars between Christians and Muslims over the Holy Land in the Middle East. These long wars were followed by famine and plague. The Black Death killed about a third of Europeans. This is when the word *quarantine* was first invented, derived from *quarantena*, which in Italian means *forty days*, the period during which all ships were required to be isolated before passengers and crew could go ashore during the Black Death plague. The *quarantena* followed the *trentino*, or thirty-day isolation period.

During this period, humans moved away from big cities. There were conflicts and schisms everywhere within the church, between the state and the church, and between revolting peasants and kingdoms. A species that was once balanced by nature metabolically and individually was now facing perennial disease, discord and disorder, all hallmarks of imbalance at individual and collective levels. Relentless wars weakened the power of church and landowners and ushered in the age of Renaissance and industrial revolution, a period that witnessed the flourishing of free thinkers, toolmakers, technologists and scientists, at least in Europe. Human-made tools became more advanced. Ships, metal works, complex weapons and printing presses gave humans unprecedented physical and intellectual reach. Widespread exchanges of knowledge through books, and goods through trade, enabled humans to enhance their numbers (quantity), mobility (speed) and reach.

As toolmakers and scientists gradually replaced priests as darlings of royal courts, human brains evolved the ability to think both in tangible (physical) and abstract (metaphysical) concepts. Scientist-engineer-artist-philosophers of the Renaissance era such as Leonardo da Vinci and Rene' Descartes helped Westerners revolutionize the ways their brains thought and conceptualized. Physical traits and balance were still somewhat adaptive and desirable as documented by *The Vitruvian Man*, a da Vinci drawing that architecturally depicts the ideal balance in proportions of a human body.

Although the Renaissance had weakened the nexus between the Royalty and Roman Catholic Church and their control over arts and culture, the 16th and 17th century scientist philosophers had to walk on eggshells when discussing matters related to the universe and celestial powers. Galileo Galilei was convicted of heresy and Giordano Bruno was burned at the stake by the verdict of the Roman inquisition (run by the Roman Catholic Church), both of them for advocating the heliocentric theory of Nicolaus Copernicus which stated the earth rotates around the sun, not the other way around. Most scientists of this era, like Descartes and da Vinci, avoided discussing heavenly celestial matters and went to great lengths not to offend the Church authorities and their large number of followers.

As the scientist philosophers had to watch the church censors, artists had to watch the Royal censors. To express public opinions about the government or kings and queens, you had to have powerful Royal patrons. If it was not because of the patronages of Henry Carey, 1st Baron Hunsdon, then the Lord Chamberlain, and later King James I, William Shakespeare could not have produced his epic plays about the Royalty with impunity. On the religious front, free thinkers such as George Fox, the founder of Religious Society of Friends, could never establish a new doctrine (Quakerism) if it were not because of the patronage of William Penn (founder of Pennsylvania) and the Lord Protector, Oliver Cromwell.

As new ideas and concepts were slowly flourishing, so were the human brains' abstraction and conceptualization circuitry. It is not a coincidence that it took until the late 17th century for an abstract concept like gravity, proposed by Isaac Newton, to receive acceptance in human brains as a real and scientific (measurable, observable) concept. Prior to Newton, there was no clear distinction in Europe between magical alchemy and science[i].

As humans brains were becoming more conceptually and cortically fit their bodies were becoming less physically fit. A brilliant human like Isaac Newton with amazing intellectual skills that gained him power and wealth, did not even pass on his genes. He was said to be a celibate and even a virgin when he died.

300 Years Ago: Adam Smith and the Emergence of *Homo economicus*

The continued expansion and exchange of new ideas, printed texts, knowledge and science, led to an exponential growth in the use of tools, trade and transportation which meant mobility of skills, products and humans. Land ownership was no longer a key to survival or dominance. You could make more money by shipping and trading a product like cotton than harvesting it. In his *Dictionnaire Philosophique*, Voltaire, one of the earliest advocates of separation of Church and State, listed serfdom under landlords and lack of mobility as the roots of human exploitation. Now craftsmen and peasants could board a cross-

---

[i] During the Golden Age of Islam from 7th to 13th century, Arab and Persian scholars such as Ibn Sina (Avicenna, founder of modern observational integrated medicine) and Al-Khwarizmi (inventor of algebra, and the namesake of algorithm) relied on evidence-based discovery in science.

Atlantic ship and head to the new world (Americas) to seek jobs and land, severing ties to their oppressive landlords in Europe.

As tools liberated humans and mobilized them, physically and socially, human labor became a traded commodity. Craftsmen such as blacksmiths (metalsmiths), carpenters, shipbuilders, seamstresses and book binders were making decent livings and had a chance for upward mobility in the new social hierarchies. Small businesses and shops grew in numbers to serve the new class of craftsmen and tradesmen. The large number of humans engaged in trade, craftsmanship and shop ownership tipped the balance of power against landlords and the Royal class and in favor of tool makers, innovators and adventurous traders. This period coincided with *enclosure* of pastures and farmlands, i.e., private appropriation of common and shared land.

As a result of mass migration of peasants, many landlords stopped agricultural farming and switched to animal farming which required less labor, raising sheep (for wool or meat) and cattle (for meat, milk or leather).

As trade and labor migration routes expanded globally, humans became massive consumers of energy. Using horse-powered carriages or plows were no longer economically advantageous (adaptive). Because coal yields a higher amount of energy per mass than wood, and because many trees were already cut by humans, coal replaced wood as the preferred source of energy. Coal mining expanded across Europe and America and employed a large number of people. The energy from coal was used in smelting and forming metal tools, which were used to build weapons and transportation vehicles, which were used in turn to explore and transport more coal. Many historians credit advances in coal mining for the expansion of the British Empire, which lasted for another century. Few other countries produced and consumed as much energy as the British did.

The energy revolution introduced by massive coal mining led to the industrial revolution of the late 18th century and an unprecedented rise in the population growth and mass production of goods and tools. Collectively, the human colony gained unprecedented quantity, reach, speed and convenience by scarring the earth (for coal and trees) and displacing or destroying other species.

Relying on more coal and tools, humans became even less metabolically efficient than their ancestors. As survival of the physically fit became survival of the economically fit, the transition of *Homo sapiens* to *Homo economicus* accelerated. The human brain had to adapt by continuing its evolutionary transition from a metabolic brain which budgeted the body and was motivated by calories, into an economically calculative brain, which was driven by dopaminergic circuits recalibrated on rewards in economic productivity. Humans started to use metrics such as GDP (Gross Domestic Product) to measure the well-being of their colony. As a result, the Industrial Revolution marks a major turning point in human history because it practically and permanently altered all aspects of daily human lives and therefore their brain.

The term *Homo economicus* was first used in the 19th century by political economists such as Charles S. Devas to criticize the economic theories of John

Stuart Mill who believed prosperity is achieved through maximizing productivity by humans who *"obtain the greatest amount of necessaries, conveniences, and luxuries, with the smallest quantity of labour and physical self-denial with which they can be obtained."*

With the focus on productivity, for the first time in about 12000 years a large number of humans became mobile again not foraging for food but for surplus economic values and jobs, as explained by the 18th century economist Adam Smith who is sometimes called The Father of Economics or The Father of Capitalism. Smith realized that it was no longer land and local serfdoms that drove the world economy and value creation. Thanks to tools and technology, mobility of humans, and specialization of crafts and trade, Smith surmised that labor and other production assets such as tools and capital (investment) were all tradable commodities with their own marketplace. Now humans could sell their skills and specializations in labor markets as easily as their goods in product markets.

Global trade and economy, as well as country-level GDP (Gross Domestic Product), a measure of *collective* economic value and wealth, immensely benefited from the productivity principle which Smith called *Division (or Specialization) of Labor*. Simply put, Adam Smith asserts that when people or nations specialize as large-scale low-cost providers of products and services traded into a global supply chain, they will maximize their wealth. Specialization arises from what David Ricardo, another 18th century economist, called a *comparative advantage* possessed by each person, nation, province or district. For example, an oil-rich nation maximizes its wealth by specializing in oil trade or converting oil to plastics at the lowest possible cost, because it has the *comparative advantage* of access to easily extractable oil reserves, whereas a populated nation with few oil reserves, specializes in low cost large volume production of labor-intensive merchandise such as garments, thanks to its large labor resources. If we need to make a wooden cabinet, a person *specialized* as a carpenter generates more *economic value per unit time* than a metalworker, but if we need a metal cabinet, a metalworker offers a better economic output per unit time than a carpenter. So if a metalworker and a carpenter trade, they could both have a nice metal and a nice wooden cabinet but if they did not specialize, each had to work hard and long hours to make a smaller or lower quality wooden or metal cabinet. So specialization and comparative advantages maximize overall economic output and standard of living. This type of interdependent global supply chain, according to Adam Smith, maximizes global profits and measures of wealth such as GDP, and ensures a world in which we can keep expanding wealth using *limited* production resources. Despite its flaws, Adam Smith's capitalism theory is now widely practiced to different degrees even in countries that were once hotbeds of centralized economies, protectionism and communism.

As labor and skills became marketable commodities of trade, new economies had to *quantify* the productive value of *unit time* for each specialization. This further removed humans and their brains from metabolic efficiency. I will share an example to illustrate. There are still small family owned dairy farmers in the

Amish region of Pennsylvania who do not use modern technology such as mobile phones and electric power or even medicine or vaccinations. They rely mainly on natural farming methods and natural health and immunity similar to their ancestors who migrated to the area in the 18th century. Many still use horses to plow their land and it takes about 5 full days to plow a 20 acre parcel (with two horses). The Amish work hard and although metabolically less efficient than hunter gatherers, are more efficient than most city dwellers. I visit the area often and have made some friends and have yet to see an overweight Amish farmer. They were reported to have been one of the first communities to reach communal (mass) natural (unvaccinated) immunity against COVID-19.

But the metabolically efficient and natural ways of the Amish lifestyle, which were rewarded and evolutionarily adaptive some 300 years ago, are not rewarding enough in the post-industrial economic world. As much as the Amish avoid hospitals, they occasionally need such interventions. An Amish farmer who binges on sugary food may need a couple of root canals by a dentist. The $2400 cost of root canal on two teeth is equal to the value of some 800 gallons of milk at wholesale price, which is about a week's worth of milk produced by a small Amish farm with 20 cows. So the modern economy assigns the same value to a week of productivity by an Amish farmer as to about two hours of a dentist's productivity. An Amish farmer eats and burns at least 25,000 kCal in that week whereas the dentist's sedentary job requires less than 250 kCal, mostly in brain and administrative work, for the few hours he spends on the root canal job. So metabolically and energetically speaking, the Amish farmer trades 25,000 kCal of labor with 250 kCal every time he needs a dentist[ii], a doctor or a lawyer. The modern economic model rewards specializations that help us *overcome* natural limits, such as science, dentistry, law, banking[iii], medicine and politics, as opposed to those that *rely on* nature such as farming or hunting.

What does all of this mean for the evolution of the human brain in the post-industrial age? It means that the human brain has to condition itself for the new economic calculus by using the same *old* conditioning circuits that evolved to budget our metabolic needs. The brain's new calculus is no longer energy (reward or expense) per unit time, but economic productivity and income per unit time of labor.

The new brain's corticotropic circuits will fear-condition us based on *conceptual* stresses and fear of *economic loss*. When conditioning us for competitions and releasing testosterone, the new brain defines a *win* as *economically* outperforming other humans and not necessarily physically fighting for food or reproductive mates. This metabolic repurposing of brain and body led to what biologist Daniel Lieberman calls *dysevolution* and to dysfunctions, imbalances and diseases he calls *mismatch* diseases.

---

ii Even after taking into account the materials and administrative costs incurred by the dentist and the cost of hay and supplies incurred by the Amish, the Amish lose a lot of time/calories in the trade

iii Banks offer financial instruments such as mortgages which are an essential *tool* in building or buying a house

The Fruit that Changed the Path of Human Evolution: Subjugation of *Homo sapiens* by *Homo economicus*

The world stage was ready for an explosion of global trade, travel, migration and colonialism. Before trains, cars and airplanes, sea routes were the fastest to travel. But there was a show stopper in sea voyages. Throughout the 18th century, disease killed more British sailors than any war did[iv]. About two-thirds of sailors would typically die in long distance voyages and scurvy was the leading cause, a disease that cause bleeding gums and prevented the healing of wounds[v]. It was not until 1747 that James Lind demonstrated that scurvy could be treated by supplementing the sailors' diet with citrus fruit. So it is not a coincidence that the British Empire colonization, world trade, and mass human migration experienced an explosive growth in the second half of the 18th century once a natural cure for scurvy was discovered. I always remind my friends and family that one could make a case that among fruits, lemon is the one that changed the course of history.

During the late 18th century, which I like to call the *post-lemon* era, colonialism spread throughout the world with help from capitalism. Wealthy merchants, members of the Royal families and government officials invested heavily in early stock corporations such as the British East India Company (BEIC), with the purpose of commercial and later military expansion into remote lands in Asia. By the end of the 18th century, thanks to the miracle of lemons, the British East India Company accounted for half of the world's trade, particularly in basic commodities such as silk, cotton, indigo dye, spices (including salt and sugar), saltpeter (potassium nitrate used as fertilizer, gunpowder or preservative), tea, and opium. Originally chartered as the "Governor and Company of Merchants of London Trading into the East-Indies", BEIC benefited from military and government support and gradually seized control of large parts of the Indian subcontinent, colonized parts of Southeast Asia and Hong Kong after the First Opium War, and maintained trading posts and colonies in the Persian Gulf Residencies. Similar companies were formed by other European nations to expand their dominance. There were several West India Companies, such as Dutch West India Company, and separately the Danish, Swedish and French West India Companies, each colonizing different parts of the world.

In the post-lemon sea exploration and colonial era, human dominance hierarchies expanded from local to global scales. Humans with *physical* prowess, such as hunter gatherers of Africa and America or farmers of Asia were subjugated by Europeans with the *economic* prowess running East or West India companies. This was the era of collusion between men of Gold and men of Guns. Advances in coal mining, tools, weapons and ship building all helped accentuate the concentration of dominance, power and wealth hierarchy in the

---

[iv] The British Royal Navy enlisted 184,899 sailors during the Seven Years War; 133,708 of these were missing or died from disease not the war!

[v] Some recent studies associate scurvy with forms of thrombocytopenia (low platelet count and dysfunctional blood clotting function)

late 18th Century. There were rebellions and revolutions as in America and France. There were also wars between the colonialists to consolidate their power.

The world had just woken up to the emergence of *Homo economicus*, an era in which large-scale, high-speed economies and tools started to replace human's physical fitness, balance and metabolic efficiency as adaptive traits. I argue that evolutionarily speaking, *Homo economicus* were a new and invasive species that can displace and conquer *Homo sapiens*, the same way *Homo sapiens* displaced the Neanderthals and caused their extinction. Neanderthals were more physically fit but less calculative and tool-dependent than *Homo sapiens*, who in turn, seem to be more physically fit but less calculative and tool-dependent than *Homo economicus*. In the new world order, even land ownership and aristocratic lineage would no longer guarantee dominance and power.

As the traditional powers of the royal and military class weakened, merchants, industrial job creators, tool makers and craftsmen gained power and challenged the ruling class. This re-balancing of power paved the way for American and French revolutions which ushered separation of government from the Church. The power of Church and royal aristocrats was severely diminished in this era. The political power in the new Republics rested with businessmen and industrialists. Most of the earlier US Presidents and statesmen were in fact successful wealthy businessmen or bankers.

As humans traded-off economic efficiency with metabolic efficiency, once more a series of plagues and diseases such as cholera decimated urban populations in congested European, American and Asian cities and trade hubs.

19th Century: Industrial Capitalism and the Reign of *Homo economicus*

Advanced tools, global trade and economic efficiency allowed capitalism to flourish. Labor became commoditized and mobile, which led to mass migration of humans, particularly from Europe to Americas, but also tragically to slavery and exploitation of human labor. The human brain, particularly in industrial countries, had evolved to assign an economic value to *any* asset including fellow humans. This type of calculus helped slave owners justify their act, even if they were religious and aware of the evil in slavery. As we shall see in chapter 6, *self-interest* often leads to *self-delusion*.

The themes of speed, quantity, reach, and convenience in human lives accelerated in the era of capitalism. Mathematically speaking, exponential (as opposed to linear) growth happens any time the rate (speed) of growth depends on the current size which is often the case with tools and technology. A fast microchip with high processing capacity can design the next generation of even faster chips at a much higher rate than an old slower one. So speed begets speed, quantity begets quantity in an exponential (non-linear) pattern.

The story of Texas Rangers and Comanche Indians is an interesting historic milestone in human evolution and the superiority of tools over physical fitness. The turn of the 19th century saw epic battles between the two groups, both known as rough and relentless warriors, over control of the Western half of the young Republic of Texas. The Comanche were known as ferocious archers who could shoot 20 arrows per minute, even when mounted on galloping mustangs. So for the early part of the 19th century, the physical fitness and archery skills of the Comanche gave them the upper hand over Rangers who relied on guns taking about 30 seconds to reload after every shot. So for every gunshot the Rangers fired they would receive 10 arrow shots from the Comanche. But there was a historic breakthrough on June 8, 1844, when a group of 15 Texas Rangers were ambushed by 75 Comanche warriors.

On that fateful day, the Rangers had a surprise in store for the Comanche. The Rangers had bought a pistol, called a repeater (or a revolver), invented and mass-produced by a Connecticut technologist businessman named Samuel Colt. The Colt pistol had a rotating cylinder and could rapid-fire five bullets at a time so it equalized the battle of shots between the Rangers, who carried two pistols each, and the physically adapted horseback archers. The adoption of mass-produced revolvers by settlers brought an end to the Comanche's reign in the area. The Comanche might have subdued other *Homo sapiens* (Indian tribes) due to better physical traits and skills, but were decimated by the industrialist economical humans. To survive in this new age, the human brains had to rewire to the adaptive powers of technology (tools) and economic scale (speed and quantity) which meant a more cortical brain.

And the early Colt pistols could not hold a candle to what came next: The 1861 invention of the Gatling gun with a rapid-firing capacity of 700-900 rounds a minute! Unlike the metabolically-driven physical limitations of natural selection, the technical and economic models were not only scalable but exponential. It took the Comanche centuries to shoot 20 arrows per minute because the mastery of archery (a natural physical skill) is limited by our muscles and body physique. Remember from chapter 3 that even under strong selection pressures, natural evolution takes time and still curbs extremes. The industrial humans, on the other hand, relying on scalable economic models, leveled up from pistols to Gatling machine guns and exactly a century later, to nuclear bombs. So the tools that initially gave *Homo sapiens* a metabolic advantage over the Neanderthals and other species, when scaled up, allowed humans to bypass their own physical limits to subdue each other.

Another tool that changed the history of *Homo sapiens* was the invention of a high speed modern cotton gin (engine) that allowed separation of cotton fibers from their seeds. The greater productivity of cotton gins (versus manual cotton separation) not only revolutionized the cotton industry in the United States, but also led initially to the growth of slavery in the American South because the demand for cotton harvesting rapidly increased. This was followed by the American Civil war and the emancipation of slaves. The triumph of industrial states in the North, specializing in mass production tools and economies of scale, over the southern (Confederate) states, specializing in labor-intensive farming and crafts, demonstrated the victory of economic over physical fitness.

Industrialist capitalists and bankers formed strong monopolies by the late 19th century, an era called the *gilded age* or less euphemistically, the *robber barons era*. Railroad, steel, coal and oil magnates were the alpha *Homo economicus* controlling an unprecedented amount of power and wealth. Millions more immigrated from Europe to America. As the skilled wages and wealth concentration grew, so did the number of extremely poor. Industrialists such as Milton Hershey owned towns, often called company towns, in which The Company controlled even grocery stores and food prices. The popular (20th century) song *Sixteen Tons*, performed by the likes of Tennessee Ernie Ford and The Weavers (Pete Seeger and his band) narrates the hopeless and grueling life of a coal miner loading tons of coal each day in one of these company towns: *"You load sixteen tons and what do you get? Another day older and deeper in debt."*

The descendants of *Homo sapiens*, who escaped the 18th century serfdom under barons and landowners, were now serving new masters, the industrialists. To escape the new serfdoms, humans had to rewire their brain again to become even more *economically* efficient. Unlike the *metabolic* efficiency that is a balancing force in nature and shaped early humans, *economic* reward is scalable. It is a never-ending quest.

20th Century: Monopolies, Nation States, Oil, Wars Followed by the Nuclear and Information Age; Science as a New Religion

The powering up of major world cities with electricity[vi], and the far-reaching spread of information through newspapers, telegraphs and telephones led to further concentration of economic power with industrial tycoons and nation states. In the US, after the Civil war (19th century) and abolition of slave labor, northern industrialists gained an upper economic hand over southern businessmen who relied on labor-intensive farming, now without the cheap slave labor. In the north, antitrust laws were passed in the aftermath of the riots against industrial monopolies of the gilded age. Nevertheless, industrialists and their families maintained a great deal of influence over politics and economy.

World cities were larger and more bustling than ever. The mobility afforded to humans by trains and cars meant information and ideas also traveled far[vii]. Unfortunately, higher collective economic output (GDP) did not mean prosperity for the masses. Several political economic ideologies emerged with solutions for fair distribution of economic gains: communism, socialism and social nationalism like Hitler's Nazi party – The word Nazi is a short pronunciation of *Nationalsozialistische*. The world witnessed two world wars, many regional wars, revolutions, riots, economic depressions and banking collapses, and pandemics such as the 1918 Spanish flu that killed or maimed

---

[vi] According to Arthur Firstenberg in *The Invisible Rainbow*, the global Asiatic flu pandemic of 1898 that killed about one out of every 1000 humans was linked to the impact of worldwide electrification projects on cellular metabolism in humans.

[vii] It is said that in the year 1913, several of the world's famous and infamous influencers lived in Vienna, Austria and were even regulars at the same coffeehouses. The list includes Adolf Hitler, Leon Trotsky, Sigmund Freud and Joseph Stalin

millions. All the while, huge economies were built by extracting every natural resource possible, including oil as the new energy source. Synthetic oil-based molecules (chemicals) such as plastics were produced; forests were decimated for logging, farming, housing and road/railroad expansions. Radio frequencies were discovered and radio towers sprouted everywhere. In *The Invisible Rainbow*, the author links the 1918 Spanish flu to the massive spread of radio towers and their impact on disturbing cellular metabolism and weakening the human immune system.

The economic selection model which rewarded scalable economic output and quantity (regardless of how it was distributed) resulted in concentration (imbalance) of economic powers so the unprecedented output (wealth) created at the collective level came at the expense of unprecedented imbalances and metabolic inefficiency at the individual level.

Amidst wars, mayhem and loss of lives, human brains were being retooled for economic efficiency, quantity and speed. Mass communication tools enabled humans not only to be more productive but also to exchange more information about how to be more productive. Remember, mathematically speaking, that type of self-reinforcing circuit leads to exponential and scale-free growth. Exponential growth in *quantity* continued particularly in the baby boom period that followed World War II. Populations increased. In the Americas, excess and surplus were everywhere.

The new economy which rewarded scaled up productivity, quantity and speed, was a perfect conditioner for the human brain's scale-free dopaminergic feedback loops, the self-reinforcing habituation and the temporal discounting discussed in chapter 4.

Perhaps the biggest change in the human brain and psyche came as a result of exploitation of science to make nuclear weapons. Although the brilliant Robert Oppenheimer, the so-called *Father of the Atomic Bomb*, later regretted his contribution, became antiwar and opposed the development of the hydrogen bomb, the proverbial cat was out of the bag. Humans saw how science could give them god-like powers so they started to religiously revere science with fanatical zeal. I am the product of that generation. I was encouraged to study math and science even by my religious elders who didn't like science's views on life and creation.

As we became more economically driven, our languages and metaphorical concepts also evolved around income and economy. We coined terms like *making a living* for drawing an income, so we subconsciously equate *living* to *income* not food or reproduction. Expressions like *bread and butter* are metaphors for sources of income. Economists use terms such as *trickle-down* to describe how tax cuts for the wealthy and their surplus money and spending, eventually *flow down* to lower income classes, like rain water that cultivates farmlands. In some languages *counting chickens*, a practice common among farmers, now refers to predicting the *economic* benefits of a plan. Our languages and brains are constantly tweaked by the concept of money.

To serve the scalable economic models, science needed to *quantitatively* understand natural patterns so it went from a holistic approach in the first part of the 20th century to a *reductionist* one in the second half. Reductionist science, by reducing whole systems into small parts, helped humans achieve unprecedented progress in terms of quantity, speed, reach and convenience and it also rewired their brains. Up until the early 20th century, even specialized humans engaged in a broad range of activities. Many carpenters were musicians and knew forestry too. Quite a few scientists were polymaths.

As humans became the main selection pressure to each other their need for collaboration and sociality decreased. Our brain, which once relied for its growth on sociality (Dunbar's number), can now rely on data, technology and analytical skills as *asocial* adaptation paths. Less prosocial trust-building activity means less of the trust-building hormone, oxytocin, which was evolution's gift to us to mitigate our anxieties and addictions, as shown in chapter 4.

Quantity often sacrifices quality. Our large personal networks today consist of social media *accounts* which are often strangers, fake profiles or bots. The *quantity* of our virtual and digital connections far exceeds Dunbar's 150 (our ancestral tribe), yet we have traded off the *quality* and depth of one-on-one human *relationships* with the *quantity* of *connections*. What does *depth* of interactions or relations mean in a world run on scale and data anyway? How can a social media corporation even measure or promote *quality* of engagements if the corporation itself is evaluated, on a daily basis, in *quantitative (scalable)* terms by a market-driven economy?

In natural evolution, asocial brains, challenged by the metabolic costs of individual learning and food collection, are on average smaller than social brains. But in the age of information technology and food excess, according to a new model[4] based on the Cultural Brain Hypothesis - which correlates brain size with group size, innovation, social learning, mating structures, and the length of the juvenile period - the cumulative impact of technology, high speed data exchange and new modes of sociality (virtual) *could* lead to an autocatalytic take-off of human brain evolution. It is too early to validate these predictions but virtual sociality and economic brains seem to certainly have weakened our physical fitness and metabolic efficiency.

Money can now buy you a lot of protection against natural risks and threats. Many wealthy people, protected by their good income or insurance coverage, build their homes on the treacherous fringes of nature like mountain vistas or ocean fronts. Our brains' risk assessment calculus is trained to assign higher risk levels to economic uncertainty than to natural threats.

The problem with the economic rewiring of our brains is that our bodies still follow the old physiology built around energy metabolism. Every time our brain motivates us to work late hours for that promotion at work, or makes us anxious to avoid a layoff, it requires sugary food for energy and it releases cortisol and glucose in our blood. But our body does not know what to do with all that extra sugar and cortisol because there is no actual running, chasing, fist fighting or tree climbing involved in a desk job.

If we could extract the risk-reward calculus that wired human brains, here is probably the shift we would see over time[viii]:

12000 years ago: It took a hunter gatherer 45 minutes of walking (about 3 miles and 300 kCal) and 15 minutes of climbing up and down a tree (another 150 kCal) to pick some fruit. The dopaminergic circuits in the brain of pre-farming humans required that the *expected value* of the fruit be at least 450 kCal to make up for the calories burnt by the forager. That is the caloric value of about 20 figs or 5 bananas. It is estimated that a forager gained and consumed about 2300 kCal during an average day which required 6 hours of hunting and foraging as compared to 4 hours of labor by a farmer.

120 years ago: The human brain still follows the body's caloric needs but by running an *economic* calculus because calories can be *purchased* now. Around 1900 AD, an average American consumed 112 grams of sugar a day, or about 450 kCal. Thanks to new tools and industrial machines such as mechanized evaporators, sugar was available at low prices of around 6 cents a pound. An average industrial age laborer made about $2.2 a day so he had to trade only 17 minutes of his labor for the economic benefit (income) to *buy* 2500 kCal worth of pure energy (sugar), as opposed to the pre-farming *metabolic* calculus of spending six hours of grueling pursuit to gather 2500 kCal.

Today: Once we lose the ability to think and motivate ourselves in metabolic terms, we become metabolically inefficient, imbalanced and diseased. In 2020, an average American consumed about 200 grams of sugar a day (152

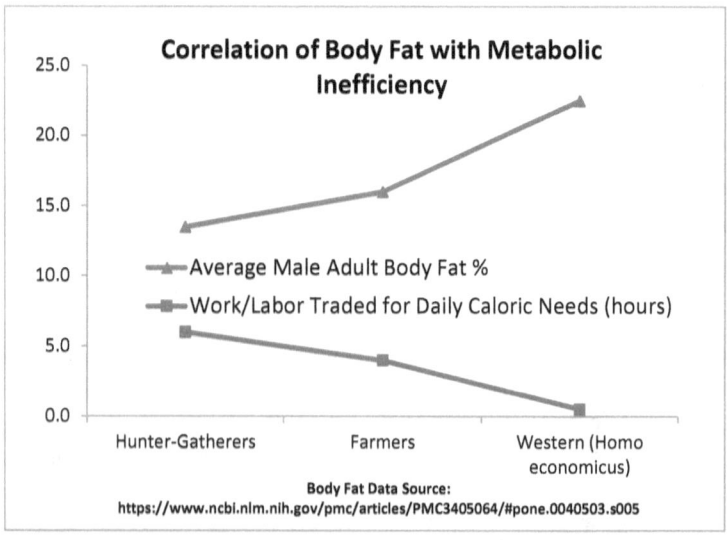

viii Other forms of energy such as fruits or fat are more costly than sugar but for the purpose of caloric cost equivalency we use sugar prices as our main benchmark. In 2020 for example, 2500 kCal secured from canned cooked pork (spam) would cost around $5 a day.

pounds a year), about twice the amount used in 1900 and one-third of all his daily caloric needs. In time-equivalent terms, this means $1.25 or about 9 minutes of unskilled labor productivity traded for the economic benefit (income) to buy 3000 kCal worth of energy.

Considering the metabolic cost of calories (sugar) throughout our evolutionary history, are we then surprised by the amount of calories we use now or by our addictions? One has to now work only 9 minutes a day and burn almost *no calories* to secure the calories needed for survival. This has upended our brain's evolutionary functions. For rewards, risks and competitive landscape, our brains now think in terms of time and *economic* productivity (money) and not time and energy (calories). It is now mostly the pursuit of income that triggers dopamine, cortisol and testosterone in our brains.

Diabetes and diseases associated with insulin and cortisol-resistance skyrocketed throughout the second half of the 20th century. It was a double whammy against our species. On the one hand, humans binged on low cost purified denatured calorie sources such as sugar and high fructose corn syrup. On the other hand, economic motivators and conceptual stressors overwhelmed our body's metabolic homeostasis. Dis-ease ensued as it always does in nature when there is imbalance and metabolic disorder.

Because of the low metabolic costs of pursuing economic goals, humans now motivate themselves on long-delayed gratification paths to collect highly rewarding *economic* fruits such as medical or law degrees or high political offices. Even if the pursuit of high economic gains is *metabolically* disastrous to us in terms of psychological stress levels, our brains often justify it because once we reach our economic goals, of say generating $50-$100 an hour, we can afford superb tools to patch ourselves up, physically and psychologically.

21st Century: The Biological and Information Wars, Altar of Science

The 21st century has seen an explosion of biological (including gene-level) and IT (information technology) solutions to our problems and diseases. Going back to the old slow life is no longer an option for most humans. Who wants to get sick and slowly heal (naturally) when we can now get back to work quickly by taking a pill or an injection? Remember the concept of *temporal discounting* in the brain's dopaminergic circuits? Once our baseline of reward expectation and prediction is upregulated, there is often no going back.

Toolmakers and rule makers still rule humans except the adaptive tools and rules of the 21st Century have changed. Currently the top rules deal with the areas of banking, law and government. Currently the top tools focus on biohacking humans (genetics, virology, medicine) and processing information.

But as new tools become new crutches, they become costly. Currently, the cost of a new model of iPhone in terms of average daily wages ranges from five days in Switzerland to 30, 50, 60 and 150 days, respectively in China, Russia, Mexico and Egypt. If you do the math based on the size of people's *disposable income* and savings account, it would take months of savings in some countries to afford a new model.

Any activity that generates an economic value is adaptive and rewarded, at least in the short term. We nicely reward peddlers of quantity, speed and convenience. Consider Amazon. If there was ever a global competition to meet the consumptive needs of new humans in terms of quantity, cost, speed of delivery, reach, and convenience, Amazon would be the winner. Yet there are several social media influencers who make a living by outraging their fans and demonizing Jeff Bezos on a daily basis, without linking his wealth to our boundless consumption levels.

The Self-Delusional Human Brains: Our Abnormal Body-Mind Connections

As discussed in chapters 3 and 4, the conservation of energy, materials and momentum (force) in nature forced our ancestors, like other species, to evolve brains that coordinated the body's metabolic budgets to reach a healthy state of homeostasis (balance).

In the process of sharing ideas (such as danger, fear or reward) and communicating with each other, humans developed the ability to *imagine* and form complex *concepts* as parcels of information that could be easily transferred and translated. Earliest concepts were perhaps related to directional space and navigation, also used by other eusocial species like bees. There is evidence that even our quantified measures of time are rooted in the concept of space[5].

Concepts could represent an abstract theme, such as cooperation, kindness, fatigue, rage, love, heaven or simply a physical entity, such as a four-legged brown wild animal called a wolf. But metabolically-speaking, the *concept* of a wolf nearby is not the same as the real (*physical*) wolf nearby because the former involves *psychological* stress, imagination and *planning* but the latter demands immediate physiological fight or flight *action*. One takes a toll on our mind and the other on our body (calorically). But over time, human brains could not easily distinguish between the *conceptual* threat of a wolf and the *visually confirmed physical* threat of a wolf because both threats activate the same fear-conditioning circuits in our brains, using cortisol and adrenaline. This is where our species separates our evolutionary path from natural evolution. Regardless of whether we face perceived (conceptual) stimuli or real (physical) stimuli, our brain triggers similar nervous reactions, sympathetic or parasympathetic.

It feels good and inspiring to realize that we are the only species that can be motivated for distant *conceptual* rewards such as a college degree, heaven, or peace on earth. But it becomes depressing once we realize our concept-driven brain is also prone to self-delusions, illusions, anxiety, depression, paranoia, addiction, mind-control, compulsive obsessions (perseverance) or psychosis[ix].

[ix] Just check out the latest volume of the Diagnostic and Statistical Manual (DSM) of Mental Disorders to see the idiosyncrasies of the human brain.

The human brain's rewiring to conflate imagined concepts with real objects is the basis of our delusionary tendencies. As already discussed, humans now spend more on illusionary characters like Mickey Mouse, Spiderman and Harry Potter, than on food. Our ability to form simulated realities inside our brains is the basis of futuristic movies like *The Matrix* and *Ready Player One*. Let's now try to understand better how the body and brain of the modern human species communicate with each other and with the world around them.

The Homeostatic Brain-Body Connections

Our body and brain work closely with each other to predict, prevent, protect, plan (the 4 P's) and decide on whether they should react in a sympathetic or parasympathetic mode, and on how to most effectively budget their metabolic resources such as glycogen, glucose, fat and protein. Our reactions all start with stimuli that could be either internal or external.

To understand how physical stimuli activate our brain circuits, it may be easier to imagine a humanoid robot that receives physical signals (visual, auditory, touch, olfactory) into a processing box (brain). Before it decides on the best metabolic path of action, the processor has to decipher both the external and the *interoceptive* (internal body) signals in terms of energy needs and availability. The processor uses the following to make sense of internal and external signals:

(A) The processor's own past (stored) *memory* of similar stimuli, matching concepts with contexts and previous experiences. This is mostly happening in the limbic brain where emotions, judgements, biases and prejudices are formed.

(B) The processor's *cognitive* algorithms and calculative decision trees on how to best serve the interest of the organism in adaptive ways (long and short term) to survive and compete against others. This is the realm of the cortical brain.

Body's Metabolic State

As discussed earlier, our body's overall orchestration and interoceptive homeostatic mode, as a whole system, is controlled by hormones as feedback signaling molecules traveling in the blood. There are four main functional homeostatic modes which can be memorized with the following mnemonic, named 4 Bs of metabolism:

(A) Break Down: Catabolic breaking down of fat (adipose tissue), glycogens and muscle, often regulated by hormones such as cortisol, adrenaline and glucagon, and metabolizing them into glucose for future cellular (mitochondrial) energy needs with the help of mediators such as insulin and acetyl carnitine;

(B) Bury: Store excess energy into fat and glycogen reserves, often regulated by levels of blood sugar and hormones such as insulin;

(C) Build: The anabolic growth of new cells and muscles and reproduction, often triggered by abundance of amino acid (protein), testosterone and estrogen hormonal signals;

(D) Burnish: Cellular repair, maintenance and healing, triggered during the parasympathetic relaxed response, sleep (induced by hormones like melatonin) and rest.

In early humans and in other species that have evolved naturally, a metabolic negotiation between the brain (deciphering the physical stimuli) and body's cellular level receptors would ultimately decide the overall mode of the body. For example, upon the sight of some new object that looks like a fruit, a primate's brain would engage its limbic (memory) and cortical (calculative) regions to assess the object's safety (as an edible fruit), its caloric density and dopaminergic reward value, and risks and metabolic costs of pursuing the food. Meanwhile, the body would assess its level of metabolic needs and resources (blood sugar levels, fat and glycogen reserves) and muscular strength (needed to chase and own the object). An honest, real-time and often quick negotiation (feedback loop) between the body and brain would ensue and result in a decision for the whole organism (body and brain) to follow. This is a simplified description of the natural metabolic process that budgets the organism's expenditures of energy in various pursuits (living) using a combination of strategies: breaking down (fat or muscle or glycogen to make more fuel - glucose available), burying (storing energy), building (growing, reproduction, and getting stronger) or burnishing (healing, sleep and rest).

In addition, the functionality levels of individual body systems (digestive, reproductive, pulmonary, cardiac, etc.,) and organs (bladder, liver, lungs, heart, intestines, etc.) are determined by whether the body and brain have decided on a sympathetic (reactive and excitatory) or parasympathetic (passive and relaxed) reaction to the stimuli. For example, if certain stimuli trigger an excited, vigilant (sympathetic) response by the body-mind axis, our sympathetic reaction would temporarily:

(A) Dilate pupils in our eyes for maximum visual sensitivity, focus and vigilance,

(B) Dilate bronchi, constriction of blood vessels, and acceleration of heart rate to accelerate flow of oxygenated blood and nutrients such as glucose to tissues and the brain (for alertness),

(C) Inhibit digestive activities to ensure energy is directed towards limbs and extremities and not stomach and digestive tract,

(D) Activate adrenal gland and release of adrenaline and norepinephrine throughout the body,

(E) Activate liver and other tissues through adrenaline and cortisol to maximize release of glucose into the blood by burning fat and glycogens (and if needed, muscles),

(F) Stimulate alpha-adrenergic (adrenaline) receptors in the urinary tract (bladder, prostate and lower distal ureter areas) and rectum to induce continence. Water is preserved to maintain blood pressure. Also, going to the bathroom in the middle of an emergency (fight or flight) is neither possible nor helpful[x],

(G) Activate sexual reproductive organs (genitalia). Explanation will follow.

Obviously, remaining in a parasympathetic state for a prolonged period leads to broken feedback loops, metabolic inefficiency and disease. The parasympathetic (relaxed, guards-down) mode is almost the exact opposite of the sympathetic response so in a natural setting, to remain homeostatic, an organism needs to *balance* the break down, build, bury and burnish cycles over intermittent sympathetic and parasympathetic episodes. The longer the parasympathetic periods are, the more we can relax and heal (burnish) and build. But how about the *bury* mode? When our distant ancestors consumed more calories in harvest or foraging seasons, their bodies turned into an anabolic mode (determined by such signals as the ratio of insulin to glucagon, as described previously) and buried (stored) some of that as fat to be burnt in colder, less abundant seasons. In modern lives, however, most humans consume sugary calorie-dense foods throughout the year so the body keeps burying that for the upcoming rainy day (season) that never comes! And the problem with fat, especially the white adipose tissue in our thighs, hips, and belly, is that it is a slow and dirty (not clean burning) fuel for the body to burn[xi].

Another serious metabolic imbalance arises from the prolonged state of sympathetic (vigilant, stressed) response because as we learned earlier, humans conflate prolonged conceptual stresses with real physical threats which are often fleeting.

When it comes to balancing sympathetic and parasympathetic states, our body's bellwether of health may be our reproductive system and genitalia, which act in an unusual way during the different phases of intimacy[6]. Initially, for genital arousal (erection in men) which is a pumping action, a *parasympathetic* state needs to dominate the body to ensure relaxation of blood vessels (and cavernosal smooth muscles inside penis). So a relaxed, low-stress state of mind is needed to achieve the arousal. Once genital arousal is achieved, a delicate and balanced coordination of neuronal circuits are essential to achieve the muscular and vascular events that lead to elongated rigid erections. But unlike erection, ejaculation is primarily a *sympathetic* phenomenon perhaps because in nature it is the climax of winning the reproductive race. Ejaculation is a highly coordinated muscular and neurological event that involves multiple sensory pathways: cerebral and spinal, integrative, autonomic, and somatic centers.

---

[x] Drugs in the family of alpha blockers, such as Tamsulosin (Flomax brand) help with urine flow in prostate and kidney stone patients by biohacking the adrenergic physiology (blocking the action of adrenaline receptors) and relaxing the body and urinary tract.

[xi] Brown adipose tissue, on the other hand, found more in women and children, is rich in mitochondria and actually burns white fat and glucose to generate energy when we are stressed by cold weather or ketogenic (low carb, high fat) diet.

So a healthy process of intimate sexual intercourse in humans starts with a relaxed state of mind and body, followed by a build up into an aroused state which uses the same aggressive circuits early human brains had evolved for competing over sex. Numerous animal studies have demonstrated that the amygdala is the pivotal structure in mediating both stress and sexual arousal and behavior. This may explain why men who are overly competitive and aggressive, like financially or athletically successful men, or people who are anxious or paranoid, experience premature ejaculation and erectile dysfunction (ED). Their overactive amygdala pushes them into a quick sympathetic mode. This condition probably resembles when early humans had a small window to reproduce before a major threat or competition arrived. But in today's world, overly competitive minds that are *habituated* to *conceptual* competitions have a hard time switching off the stressful fight or flight mode.

The health of the reproductive system and love-making experience will be introduced in Chapter 8 as one of the key indicators of the state of balance between our sympathetic and parasympathetic reactions and our overall homeostatic balance and health.

As hunter gatherers we evolved in almost clockwork fashion in order to stay in tune with our bodies' metabolic needs. This is how we evolved our circadian rhythm as a prediction algorithm, based on the timing of the external (light) stimuli, the best time to crank up vigilance (cortisol release), and the best time to slow down to rest and heal.

The Non-Metabolic Limbic Brain: Emotions and Concepts

Our limbic system is where we consolidate sensory signals (such as external odors or internal pains) and associate them with episodic (as in amygdala) or contextual (as in hippocampus) memories. And some of this memory can be passed on to us from our progenitors. So if in our memory (experience) we have associated the stench of a foul smell with rotten or toxic food, we develop an emotional (affectual) *disgust response each* time we receive that sensory stimulus. Many humans inherently fear darkness perhaps because darkness meant insecurity and risk during most of our evolution. We often welcome the smell of mild burning firewood maybe because in our evolutionary memory that means food, heat and safety are nearby. So our emotions are basically *constructed* based on our past (near or distant) memory experience.

Because each one of us has a unique set of experiences, our emotional spectrum is also unique to us. For example, when I was growing up, I vividly remember my grandfather making a disgusted face whenever he smelled raw onions. For years, even after I grew up, I associated raw onions with a sense of disgust without realizing the *affectual* root of my *displeasure*.

The uniqueness of humans is based on our contextual memory of abstract concepts. The luxury to imagine threats or rewards that are not metabolically relevant to us is probably rooted in the advent of concept-driven languages and farming. Thanks to grain silos and domesticated livestock, humans gained surpluses and used less of their own muscles, which meant they had the leisure and luxury to stop worrying about metabolic frugality. With more free time and excess calories, humans had the luxury to think and conceptualize in metabolically inefficient ways. For example, if a farmer envisioned and conceptualized a totem that would protect his family from natural threats, he could afford to offer some of the grain surplus or sacrifice some of the livestock for the totem or for the person representing that totem. The number of shared concepts, ideals and words grew as humans grew their surpluses and sizes of their families, communities and social networks.

Shared wisdom (science) helped humans advance their collective welfare and widespread concepts such as imaginary figures and epic icons gave humans something to talk about and strengthen their shared social bonds and values. In today's world, some international fan clubs of movies, actors, sports clubs and athletes have more zealous followers than ancient religions.

In our quest to understand the complex world, we developed a unique ability for *abstraction*, a process in which the brain compresses and integrates our sensory input into functional (not physical) concepts such as metaphors, symbols, ideas, ideals, ideologies, themes, that will impose meaning on our life and the world surrounding or inside us. The problem with abstraction is that when it conflicts with the real physical input (feedback) from the outside world or from inside our bodies, we often stick to our constructed (abstract) perception. This is why we can be called a *self-delusional* species. We have learned to survive with *simulated* brain concepts and ignore our body's and nature's physical feedback.

Conceptual thinking uncoupled our mind-body axis and coordination. Our limbic brain evolved to categorize sensory signals into all sorts of sensations (affects) based on complex concepts. Based on those concepts, we developed symbols, metaphors, analogies, parables, similes, poetic and figurative speeches that help us understand *implicit* meanings. For example, we understand that Kafka's *Metamorphosis* isn't really about a cockroach, or *soma* in Huxley's *Brave New World* is more than just a happiness pill, or Ionesco's *Rhinoceros* is not really about a town overrun by real-life rhinoceroses.

We may understand metaphorical nuances but value-based social concepts become very real to us. These include pride, shame, honor, envy, nationalism, religion and the most recent and strongest of all, money, a concept which was not prominent in human lives until scalable economies and capitalism wiped out local barter mechanisms in favor of global trade and fiat currency.

A concept-driven brain forms judgments based on *gut feelings*, often a net effect of both internal and external senses and conceptual and social constructs. Research by psychologist Simone Schnall and her team[7] shows that foul smells and repugnant scenes can not only evoke a physical sense of disgust, it can also evoke a disgusted *affect* (gut feeling) and a punitive moral judgment. People who were sitting in a foul-smelling room or at a desk cluttered with dirty food containers judged questionable acts like lying on a résumé or keeping a wallet found on the street as more immoral than individuals who were asked to make the same judgments in a clean environment. The research indicated the importance and specificity of *gut feelings* in moral judgments. This *affective morality* is unique to a concept-driven brain. It has been demonstrated that regardless of whether we feel physically, socially or morally repugnant (disgusted), similar parts of our brain are activated.

As mentioned earlier, one part of the human brain which is shown to be engaged in our gut feelings is the insula. The emotion of disgust is apparently evolved to protect us from pathogens, and elicited by substances like feces, vomit and putrid food. The insula is closely related to disgust because it serves both gustatory and visceral motor functions including the control of vomiting. In modern concept-driven human brains, the insula is activated by a broad range of disgust-related stimuli such as disgusted facial expressions, unpleasant odors, pictures of rotten food, and unfair acts. Some research indicates that the insula plays an important role in the experience of not only unpleasant but also pleasant bodily feelings. In brief, the insula seems to be involved in the emotional perception of bodily feelings. If something smells fishy, literally or metaphorically, the insula is the part of our brain that orders us to avoid it.

Our brains are so ambivalent about real and metaphorical concepts that subtle physical cues about cleanliness can affect our socio-political judgments. In a psychosomatic experiment[8], participants who were randomly positioned in front of a hand sanitizer gave more clean-cut conservative responses to a survey about their moral, socio political and fiscal attitudes than individuals assigned to complete the questionnaire at the other end of the hallway.

The intertwining of conceptual and physical disgust in humans has real implications. In 2010 a gubernatorial primary candidate in the state of New York mailed out thousands of campaign ads impregnated with the smell of rotting garbage and photos of his rival, the scandal-tainted Governor.

As deliberative as we think we are, our opinions and feelings are often quite visceral. That is why many people vote for a political candidate or buy a product that makes them *feel* better or *seems* more reliable. The next time you watch TV and video commercials for food products, cars, pharmaceuticals or politicians, try to dissociate your feelings and act as if you were paid to be a harsh analytical critic of the ads to see how they are loaded with visceral triggers.

So human emotions and affects are basically our own constructs in a world we simulate to confer a meaning to various stimuli. Apart from disgust, our brain constructs other affects such as warm and cozy feelings. Yale University psychologists have shown[9] that people judged others to be more generous and caring if they had just held a *warm* cup of coffee, and less so if they had held a cup of *iced* coffee. In a second study, they showed people are more likely to give something to others if they had just held something warm and more likely take something for themselves if they held something cold. These observations are more evidence that our brains have a hard time separating physical and conceptual (emotional) affects. This is the root of the *self-delusional* brain in humans.

Our decisions are driven by all sorts of affects such as shame, guilt, greed (money), envy, vengeance, honor, etc. We are also prone to making moral judgments based on *gut-feelings*, even those among us whom we elect as the most rational and judicious, i.e., politicians and judges. Researchers at Ben Gurion University in Israel and Columbia University examined decisions by eight Israeli judges who ruled on convicts' parole requests. Judges granted 65 percent of requests they heard at the beginning of the day's session and almost none at the end. Right after a snack break, approvals jumped right back to 65 percent! The authors attribute this effect to the fatigue and hunger experienced late in the day by judges, i.e., metabolic deficiency in their bodies and brain. Apparently fatigue, similar to anxiety, pushes human brains into a perseverative (autopilot) mode, avoiding difficult cortical decisions and distinctions.

We also tend to make our affectual judgments *unconsciously*. This is a realm that Carl Jung calls a *complex*, a hidden but *real* part of our personality. Because humans are all driven by self-delusional unconscious complexes, they may not always be aware (conscious) of their *intent*. This is why we should resist assigning *evil intent* to others when they promote a defective product or praise a damaged person. People often exaggerate or obfuscate facts *unconsciously* when motivated by personal rewards and gains such as bonuses, publications, beach houses, contracts, pride, fame, etc. In the world of *Homo economicus*, whistleblowing and straight talk can mean loss of economic value. When a modern highly paid scientist *thinks about* rocking the corporate boat, he or she activates the same brain circuits that fear-conditioned our hunter gatherer ancestors when they saw a tiger or a grizzly bear. In the modern world, fame and fortune may be conceptual but their rewarding impacts on our motivations are as real as the corn or wheat crop were to the farmer, and fruit and game meat were to the forager.

Our concept-driven brains, managed by limbic judgments and cortical calculations (often economical) are now the dominant voice in the body-mind communications which were once ruled by the body's metabolic needs. Humans can be called a brain-dominant organism.

Brain's Cognitive Cortical Algorithms

The cortical human brain, originally evolved as the executive metabolic manager of body functions, is now a master of mind games. Cynical evolutionary biologists attribute the large cortical brain in humans to their ability to lie and play mental games as an adaptive trait in response to selection pressure by other self-serving, cunning, self-delusional humans. According to cynics, the Dunbar's number reflects shrewd competition among collaborating but cunning humans that have to constantly outsmart each other.

And the cynics may have a point. Unlike humans, animals do not use complex speech or conceptual communications so they are unable to easily lie and manipulate each other[xii]. For example, a dog's anal glands release distinct pheromones that reveal his (her) health, age and sex. If weak, sick, or not sexually virile, a dog cannot do much to cover up their condition other than tucking in their tails to block release of anal pheromones. Nature is often unmasked and communicates in real time.

The ability to conceptualize, calculate, and use languages and symbols such as clothing, cars and titles, to mask our biological states and intentions, has placed us on a convoluted evolutionary path. Humans wear multiple personas that make it near impossible for them to exchange the kind of real-time feedback that is responsible for symbiotic balance in nature. Because humans are game players (mind games that is, complex mathematical simulations are now used to predict adaptive social behavior in humans. This is the field of *game theory*.

The Ultimate Concept: Economic Output (Money)

Today, at least in the industrial world, the concept of money (economic output per unit time) is central to how our limbic and cortical brains are conditioned. Unlike food which is calorically beneficial as a reward, money has no direct metabolic or physical benefits so its value is strongly linked to human-made economic parameters such as inflation, exchange and interest rates. For *Homo economicus* brain, economic productivity (income) is the key factor to optimize in decision tree algorithms. Many of us (subconsciously) calculate the impact of any decision in life in terms of its impact on our short- or long-term material assets or safety nets. Conceptual values such as nationalism, friendship, loyalty and ideology (religion) still play a role but our brains are conditioned not to ignore the economic aspect of our decisions and behavior. I have seen way too many fair-weather friends, patriots, ideologues, self-censoring journalists and humanitarians who are mainly driven by economic gains and nothing else.

The concepts of income and money, unlike *qualitative experiential* values such as integrity or friendship, are *quantitatively* measurable so they are strong activators of our goal-oriented motivational dopaminergic circuits which can count dollars as easily as calories. Quantitative economic concepts also heavily engage our *cognitive cortical* (calculative) brain. So the brain of a modern human, particularly *Homo economicus*, is strongly driven by economic risk-reward calculus for the unit time spent on any pursuit. Our brains are now efficient *numerical computational* machines. Research shows[10] out of the 20 watts of glucose available to the brain (equivalent to a small light bulb), cortical *computation* consumes only 0.1 watts of ATP, indicating the strong *quantitative* conditioning of modern human brains.

Consider a *Homo economicus* is invited to a fundraiser for a charitable cause that would cost 4 hours of his/her time. If the person's usual economic output

---

[xii] Some animals do use simple physical bluffing and camouflaging strategies. Grey squirrels, for instance, have been observed to dig fake holes for storing nuts as a diversion and subterfuge when there are other squirrels in the vicinity.

is $250 per hour, s/he would only participate if the indirect benefits (such as networking with power brokers or PR gain through virtue signaling) is worth the $1000 in his/her *opportunity cost*. If the economic opportunity cost is $5000 an hour, s/he would expect gains of at least $20000 in return. I personally know many well-meaning professionals whose brains are wired like this.

Today, even many of our value judgments are shaped by economic concepts and contexts. For example, if a human injures another in a street fight, we often imprison the attacker but if a human injures another in a boxing arena or in a war zone, we may cheer for it and even reward it because sports and wars can create jobs and stimulate the economy (GDP).

In the age of *Homo economicus*, multinational private corporations will have as much power as, if not more power than, national governments in shaping our future. Recently, the World Economic Forum (WEF) and the United Nations (UN) signed a memorandum of understanding to partner with each other in matters of global governance including technological, health, environmental and educational policies.

As I write this book, the human brain is amidst another transition to replace old world's metaphysical concepts with a single fiat virtual concept called money (economic output) and soon with digital currencies and tokens, which are not even tangible or fungible. I have seen too many families and friendships break apart over money. I saw too many back stabbings in my corporate years, not over basic food or shelter, but over six-figure salaries, yachts and vacation homes. Humans are now in a race to level up their economic game and scale up their cars, homes and income. We are all becoming *Homo economicus*, which is the topic of my upcoming book. As will be discussed in chapter 7, our judicial, political and even educational and medical systems are now mostly bogged down with issues related to the distribution of money. Money, as the supreme concept invented by *Homo economicus* to facilitate *quantitative scale up*, when applied to our problems, scales them up too! Money is our main tool for problem solving. Remember Maslow's witty quip *"If the only tool you have is a hammer, you tend to see every problem as a nail!"*[11].

The Economic vs. Metabolic Brain

In evolution, metamorphosis is an after-birth biological process which causes relatively abrupt changes in the animal's anatomy and functions. I use the term only loosely to refer to our transition as humans from a somatic metabolically efficient being at birth to a cerebral calculative being, or what psychiatrist George Engel calls a *Biopsychosocial species*. Thanks to a brain driven by scalable economic outputs, we have diverged our path from principles of natural evolution and selection outlined in chapter 3. Here are some differences:

Humans Evolve Memes and Behaviors

Many evolutionary biologists believe *memes* are as powerful as genes when it comes to human evolution. Memes are ideas, behavior and cultural themes that human brains select to spread over space and time, like species do with genes and viruses. The list of powerful universal memes today includes consumerism, vanity, fashion, religion. In humans, cultural evolution seems to serve the genetic evolution.

Non-Physical, Cerebral Adaptation

Unlike in natural evolution, physical features are no longer the main adaptive traits in modern humans. Some of the fastest running human sprinters are among the poorest and live in Africa or in countries like Jamaica, serving some of the most physically unfit *Homo economicus*.

The adaptive advantage of a calculative brain in modern times is somewhat balanced by the maladaptive dependence on gadgets. In contrast to natural ecosystems, controlled environments lead to shrinking of the cortical brain over time if the organism's survival depends on easy-to-use crutches and gadgets. Studies have shown that wild rats *challenged* in nature are more resilient and better problem solvers than lab rats *trained* with advanced gadgets[12]. In times of uncertainty, the novel and random nature of selection pressures seem to select for a more judicious, nimble, and creative cortical brain over a reactive brain that relies heavily on amygdala and hippocampus. This is the Achilles heel of some of the most calculative humans (*Homo economicus*): Uncertain unpredictable times not manageable by gadgets, expensive tools and crutches. I once knew a young man who could not drive a short distance from work to home without using a GPS, even though he traveled that route every day.

Metabolic Inefficiency

Metabolic efficiency is replaced by economic efficiency which is not naturally self-balancing. In modern economies, wages and money supplies are not even directly linked to real productivity so economies are fully scalable: The sky's the limit. But nature never works like that. The sky's *NOT* the limit for the length of a giraffe's neck. It is evolutionarily limited by the height of leaves and buds on local trees, which are in turn kept in check by other shoots, branches and trees competing for sunlight and soil nutrition.

As a metabolically inefficient species, we have evolved high basal metabolic rates - the number of calories our body needs for basic functioning. Multivariate regression studies[13] taking into account body size and physical activity have shown that the total energy expenditure in human bodies exceeds that of chimpanzees and bonobos, gorillas and orangutans by approximately 400, 635 and 820 kcal a day, or in percentage terms, about 27%, 33% and 50%, respectively[xiii]. This is reflected in the humans' greater basal metabolic rate even at the organ level. The paper concludes that our high metabolism seems to have evolved independent of the expected energetic trade-offs of larger brains or higher reproductive outputs. I believe this conclusion confirms that concept-

---

[xiii] Amphibians and reptiles, as a consequence of ectothermy and slow anaerobic resting metabolic rates, consume significantly less energy than endotherms. For example, a free-ranging fence lizard, consumes only 3-4% as much energy per day as a bird or mammal (normalized by weight).

driven human brains evolved to metabolically decouple from their original hominoid body. According to researchers, to manage their high basal metabolism, pre-industrial humans evolved behavioral and anatomical adaptations such as energy-efficient locomotion (bipedal walking), dietary changes, food sharing and maternal provisioning to mitigate the risk of energy shortfalls.

A greater need for energy also means a greater need for storing it as fat. The average percentages of body fat in human males and females, at 22.9% and 41.1%, respectively, are much higher than the body fat in male and female gorillas at 15.2% and 13.9%, respectively[xiv]. The human's so-called *Thrifty Genes* (storing fat) became problematic when the industrial age ushered in easy cheap calorie sources that triggered cortisol and insulin overloads. Currently, some 20-40% of people in some industrial nations are considered obese, defined as a body mass index (BMI) of greater than 30.

Temporally and Spatially Limitless

Unlike in naturally evolved species, *current* time and space do not limit adaptation and decisions in modern humans. A concept-driven brain is often dwelling in the past or future because concepts, whether rewarding or painful, are not limited to current time and space. Humans have mastered the art of planning and pondering retirement, a long time in the future and in a remote location from their present spot. Being present and basking in the moment's sensations is such a challenge for humans that we now have countless books, gurus, and even cults dedicated to guiding us into being *present*. The meditation industry may not be as large as the imagination industry (entertainment, gaming, etc.) but it is certainly experiencing an exponential growth in recent years because for many of us a meditative state resets our brain and allows us to become even more *economically productive*!

Quantity and Speed vs. Quality and Moderation

In nature, quality, for example of a genetic lineage, is the ultimate goal of evolution and means resilience and passing the test of time. Nature uses time (patience) to perfect *quality*. The modern human's evolution, on the other hand, is driven by economic (and not metabolic) selection pressures so it rewards *quantity* (scale), speed and unfettered growth. But this can create a conflict and cognitive dissonance in our species because while our market economies, driven by commodification and commoditization, condition us to think in terms of "*exchange values*" of time and resources, humans still sensually and emotionally (spiritually) enjoy what economist Yanis Varoufakis calls *experiential values* in life, which cannot be commodified (quantitatively measurable, scalable, comparable).

Also consider the effect of *time* on the drugs and chemicals we have developed with a *short-term* focus to provide a speedy solution to our problems.

[xiv] When it comes to percent body fat among primates and hominoids only humans exhibit a significant sex difference, with males having less body fat than females.

Chemicals break down over time or react with each other or with oxygen or water. Pharmaceuticals turn into metabolites that are further metabolized into secondary metabolites by different organs over time. We now know that residuals of many drugs, artificial pesticides, chemicals and dyes, as well as heavy metals and toxic substances that our body cannot metabolize[14], accumulate in the fatty myelin sheaths (Schwann cells) protecting our neurons. This may cause demyelination, toxicity, loss of neurons (neuropathy), and encephalopathy (inflammation in the brain). Yet, most chemical or pharmacokinetic studies do not track long term impacts in the entire human body, in the ecosystem, or in other animals or humans[15]. I will discuss the problems of reductionist science in chapter 7.

Speed begets speed, and quantity begets quantity. Remember the mathematical formula that predicts exponential growth if the rate of growth of a phenomenon depends on the size (quantity) of the phenomenon. It is now demonstrated that the rate of technological changes such as computational power of computer processors, is accelerating in an exponential fashion. In 1938, architect and inventor Buckminster Fuller even introduced a word to describe how humans are constantly *doing more with less* in science and technology: *Ephemeralization.*

Unlike in natural evolution, moderation and balance are not rewarded in economic terms. Excess and imbalance are not maladaptive traits. If a human has an equivalent of a peacock's tail in terms of ostentatious vanity and wealth, he or she could easily hire bodyguards, lawyers or security systems to protect against potential predators.

Excess may not be economically maladaptive but it is stressful to maintain. This is evidenced by the high frequency of self-inflicted harm or suicide among alpha humans. The movie review website IMDB has a page dedicated to *Actors and Celebrities Who Committed Suicide*, usually updated once a week.

Parasitic Monocultures vs. Symbiotic Biodiversity

A species driven by scalable numbers is often blindsided with long-term consequences of its decisions on other species or even its own. The reason lichens, fungi and trees live in symbiogenesis is because they metabolically need each other, locally. But humans rely on their income and not on *local* nature for survival. Bill Gates for example, does not metabolically or even economically rely on the thousands of trees (wood logs) cut (outside his home area) to build his residence, or on the species once housed in those trees. I am not being facetious or judgmental. I am holding up a metaphorical mirror to all of us, reflecting on the fact that our environmental issues started when our species stopped relying on *local* symbiogenesis and metabolic efficiency for survival and progress.

Symbiosis often requires that animals adapt to their ecosystem. Colors evolve in butterflies in harmony with colors of local flowers and vegetation in response to predatory selection pressures. Modern humans, on the other hand, often go to great lengths and great expenses to change their ecosystem to fit their own taste, free from the limits of symbiotic and biodiverse evolution. Thanks to innovative tools, humans can now use a large amount of resources (pesticides, lawn mowing, and water) to build and maintain green colors even on a backdrop consisting of dry deserts. That is how the green fairways of Coyote Springs Golf Course in Moapa, Nevada - which cost $40 million to build - and the many green fairways in Palm Springs, California (with 300 days of sunshine each year) became possible. Unlike the butterfly which camouflages to survive, humans change nature's colors (greens for golf) to suit our taste. We have become the selection pressure to many other species.

But our actions are often ecologically *myopic*. Take the case of our transition of forestlands into population centers in the past two centuries. First we cleared and demolished a balanced biodiverse ecosystem of meadows and old growth forests to build structures (buildings, roads, factories, golf courses, farms) using materials from other demolished ecosystems. Then we realized the imbalance we had caused so we went back and *unnaturally* forced monocultures (single crops or trees) into lands which were often not natural habitats for the tree or the crop species. But unlike biodiverse ecosystems, monocultures are not biologically resilient so we then resorted to chemicals, pesticides, irrigation and costly human interventions to keep the cultivations alive. But our buildings, roads and driveways had compacted the soil and weakened the roots of plants and trees while the chemicals and pesticides we used hurt the symbiotic mycorrhiza (naturally diverse soil fungi) that help tree roots. Weaker roots meant our plants and crops withered easily particularly because the soils were depleted of nutrients after commercial exploitation. So the short-sighted humans decided to rig and outsmart nature once more by (over)using chemical fertilizers to force the plant to perk up and become more *productive*.. Farmers today apply 5-15 kilograms of phosphorus (in fertilizers) for each kilogram of phosphorus available in crops we consume as food. The excess phosphorus (4-14 times the amount we consume) builds up in the soil and in runoffs into waterways, causing algae bloom, green scum and toxins that foul drinking water and suck up dissolved oxygen. This type of *imbalance* kills fish and other aquatic life. So to fix that problem, humans found an ingenious solution. Who needs a clean lake or river when we can raise farm-raised fish? Farm raised fish is now big business and loved by *Homo economicus* because it is economically scalable and contributes to GDP and job growth independent of how toxic our natural waterways become. So we have achieved scalable economic growth at the expense of sustainable biodiversity and natural symbiosis.

And once humans invent a tool, we repurpose it often and find new applications. I worked for many years as an applications development scientist. My job was to find new applications for existing technologies and materials. For example, synthetic fertilizers use oxidizing nitrogen-containing chemicals such as nitrates and perchlorates, which have found applications ranging from solid rocket fuels (propellants) to medicine, matches, pyrotechnics (flares, smoke generators, tracers, incendiary delays, fuses, photo-flash compounds, fireworks), cattle feed, magnesium batteries, and as a component of automobile airbag inflators.

I call modern humans *myopic* because we *literally* are short-sighted. We only care about the air-conditioned, walled boxes we keep clean and protected but forget that our *real* habitat is the ecosystem beyond our immediate bubble. Consider our neighborhood which is located near 2 major rivers and several creeks. Because of contamination, we cannot rely on any of the water sources for drinking or fishing. Most of my neighbors who used to rely on well water now buy water in plastic containers bottled long ago in a factory far away. And they buy fish raised weeks earlier in a distant commercial fish farm. In nature, usually only parasites or invasive species violate symbiotic biodiversity and upset the balance in a flourishing local ecosystem.

From Scale-Free Bottom-Up to Scalable, Top-Down

As we saw earlier, in nature, locally driven adaptation leads to bottom-up scale-free evolution and order. Human economies, however, gravitate towards top-heavy systems because larger scales are economically more rewarding than smaller ones. So don't expect to come across Fibonacci, fractal type or Turing's reaction-diffusion type patterns in human-made structures. Large-scale metropolitan cities with 3D structures and large population densities look grossly different from small towns and villages with 2D structures and flatter buildings. Organizational charts, tax codes, and revenue per employee in large corporations are often nothing like those in a small family business or restaurant.

Corporate-dominated globalism is not as much an evil conspiracy as it is an *outcome* of the competitive superiority of large-scale businesses over smaller local and national economies. Everything else, our societies, politics (human relations), family, national, ethical, moral and religious values are now all closely intertwined with an scalable economic engine that feeds most of us. There are very few independent communities of farmers (like the Amish) or foragers (the *Hadza* of Tanzania) left in the world. Everyone's income is somehow linked to a large and concentrated economic system of money creation, distribution and regulation managed by banks, corporations and governments (politicians)[xv].

There are studies that try to apply metabolic efficiency (flux and conversion of materials and energy) to model civil structures and human cities[16], or human economies, environments and technologies[17]. Nevertheless, the artificial boundary and initial conditions used to model a man-made system will prevent it from being part of a large scale-free ecosystem. If anything, human-made cities and colonies are grossly imbalanced as compared to natural systems. In large metropolitan areas, there are skyscrapers in close proximity to broken dilapidated buildings, ghettos and homeless camps. In Echo Park, Los Angeles,

[xv] The COVID-19 pandemic seems to be upsetting the scalable global supply chain models, which may encourage people to adopt local, resilient, small scale community-based economies like the Amish.

for example, you can find homeless camps, million-dollar homes, and one bedroom apartments that house large families, all within a few miles of each other. If human cities were organs of a larger colony, you could liken them to cancerous growth, a small dense fast-growing cluster of cells monopolizing resources away from a large number of normal healthy cells. Compare the road maps in major metropolitan cities to scanning electron micrographs of angiogenesis in tumor cells to see the resemblance. Also, remember what the humanized virtual Agent Smith said in *The Matrix* (1999 motion picture):

"Every mammal on this planet instinctively develops a natural equilibrium with the surrounding environment but you humans do not. You move to an area and you multiply and multiply until every natural resource is consumed and the only way you can survive is to spread to another area. There is another organism on this planet that follows the same pattern. Do you know what it is? A virus. Human beings are a disease, a cancer of this planet."

Overriding Adaptive and Epigenetic Memory

Natural evolution has a memory and develops robustness through epigenetics and adaptive immune systems. But in humans, it appears that we are now accumulating *maladaptive* epigenetic *defects* because we are constantly under allostatic metabolic loads (biological stresses associated with chronic deviation from homeostasis). Examples include shrinkage of dendritic neurons in chronically-stressed brains, chronic allergies, immunosuppression, and downregulation of our cellular receptors leading to disorders such as insulin resistance and glucocorticoid resistance. To override these epigenetic disorders we often *biohack* our body's natural pathways like the AMPK pathway that controls our cellular metabolism.

Individualism

Natural evolution works at individual as well as group level through altruism, kin selection, group selection and swarm intelligence. Eusociality became an adaptive trait because it greatly reduced metabolic costs of survival for individuals. The brain of a modern human, however, is now strongly conditioned to maximize his or her *individual* economic productivity. As a result, when hard times hit, as during the Covid-19 pandemic, we see a high level of discord and division between friends, family members (kin), countries and even otherwise altruistic humans.

Contrast our pandemic panic to group-level anti-pathogenic defenses, collectively termed *social immunity*, in other eusocial species. Social insects like ants, for example, can detect the immune status of their nest-mates and if they find them infectious, instead of avoiding them, often respond with *caring* strategies such as mutual grooming (allogrooming), waste management, disinfection of the nest with antimicrobial substances and global patterns of social interactions that reduce disease transmission (organizational immunity). Allogrooming is a widespread behavior in social insects that involves removing infectious particles from, or spreading antiseptic substances onto, the body surface of nest-mates, this behavior can increase the survival of pathogen-exposed individuals[18].

But surely we think we are smarter and more individualistic than ants so what can evolution teach us about other smart species that have become individualistic? Evolutionarily speaking, we might be following the path taken by cephalopods.

Literally meaning brain (*cephalo*) in feet (*pod*), cephalopods (like octopus) are masters of disguise, camouflage and escape. After their ancestors lost their protective shells, cephalopods evolved complex problem solving brains, distributed in their limbs, which adapted them to predatory selection pressures. But cephalopods are *individualistic* problem solvers so despite their complex predator-avoidance intelligence, they live short lives. Solitary (individualistic) problem solving means cephalopods have to bear the metabolic cost of stressful change alone without any long lasting social bonds (as in eusocial species) that act as safety nets against predatory conditions.

The Body-Mind Axis: The Allostatic Roots of Disease

The human brain initially evolved to help with energy budgeting of the body so it was the body which was in the driver seat. The brain of a modern human, however, is in the driver seat of the body-mind axis because it makes decisions based not only on the body's physical sensory inputs but also on two other *perceived* inputs:

(A)   The brain's own limbic system for judgments, affectual emotions, experiences, prejudices, gut feelings and opinions, all shaped by concepts (particularly the economic concepts);

(B)   Environmental psychological signals. That is why *hearing* the *news* of losing a job could elicit as strong of a stress reaction as facing a serious physical threat. That is why staunch followers of sports, religions and politics, when feeling insulted, may experience the same high blood pressure and sympathetic nervous response that our ancestors experienced when attacked by a predator.

The cerebral process of perceiving stress and reacting to it is depicted in the following diagram. Exposed to a long list of environmental, prejudicial and psychological stressors, modern humans experience prolonged periods of homeostatic imbalance (allostasis). This leads to disruptive (sympathetic) physiological responses such as disorders in blood pressure, glucose, insulin and cortisol levels, digestion, sleep patterns and libido. Our brain has also evolved an adaptive mechanism to minimize the metabolic impact of long term trauma on our bodies. It is called *neuroplasticity*.

Neuroplasticity is the brain's dynamic adaptive ability to grow or shrink neurons (synapses, dendrites, spikes) in response to rewards, stressors and trauma. Firing of neurons is a metabolically expensive process. By some accounts, most energy used in the brain is spent on reversing the ion flux through postsynaptic receptors, which consumes 50% of the signaling energy[19]. So during exposure to prolonged excitatory (sympathetic) stimuli, on energetic grounds, fewer or smaller receptors and spines would be preferred as they consume less energy[xvi]. This type of energy dynamics over time leads to shrinkage or expansion of dendrites (neurons' receiving tentacles similar to tree roots) and dendritic spines in response to stress overloads. On the other hand, if our survival depends on pursuing certain rewards and learning certain routines, neurons would branch off or grow spines and receptors to allocate more energy to that process, although too much dopamine will exact a metabolic toll on the brain too.

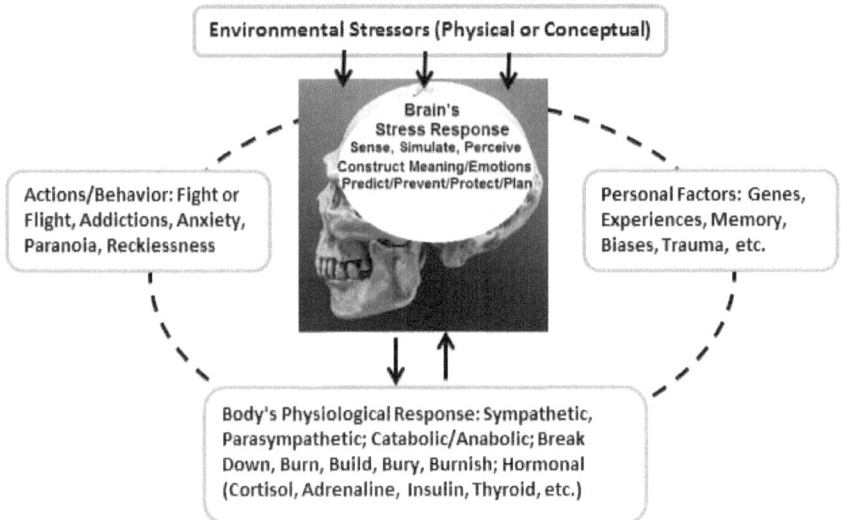

In chapter 2 we discussed how our brain's neurons, like other cells, use adaptive strategies such as arborization (tree-like branching) to achieve long-term metabolic efficiency. Like tree branches, axons and dendrites can sprout new buds and branches and retract old ones. If you are a musician and practice a certain instrument, your brain's neuroplasticity expands the auditory cortex devoted to the sound of that instrument. If you are an old-fashioned taxi driver not using GPS or even maps, your brain tends to form a complex spatial map of streets and buildings in the city which tends to enlarge the part of the brain responsible for memorizing maps, the hippocampus[20]. But this sharpening of memory would not happen for a driver using a GPS as an indispensable tool (crutch). GPS-dependent drivers may instead be expanding the part of their brain responsible for finger-eye coordination (as they constantly text or type),

---

xvi For excitatory synapses to be able to repeatedly transmit information on a time scale of milliseconds, the diameter of synaptic boutons and spines must be less than 1 micron, to allow rapid glutamate clearance by diffusion to glutamate transporters in surrounding astrocytes.

the cerebellum. Neuronal plasticity helps us change and adapt as a result of learning and life experience. It basically means our experiences in life can expand or shrink certain parts of our brain to meet our temporary needs while maintaining neuronal homeostasis.

Neuronal plasticity is the reason behind the inverted U shaped (bell-shape) dose-response relationship between cortisol levels and cognitive sharpness/memory recall performance, as discussed in chapter 4. When stress levels become too high or too frequent, mental sharpness (memory and vigilance) and physical performance decrease. The inverted U is also aligned with the general concept of *hormesis*, which is the adaptive response of cells and organisms to moderate (usually intermittent) stresses and their inhibitory toxic response to prolonged, high frequency or acute (high dose) stimuli.

Prolonged stress or PTSD will expose neurons in the frontal cortex, the hippocampus and the amygdala, all containing numerous glucocorticoid receptors, to excitatory (glutamatergic) surges of energy. This can lead to stress-induced dysregulation of glutamate release which in turn results in the shrinking of dendrites in the hippocampus and frontal cortex and expansion of neurons in the amygdala. Furthermore, PTSD is shown to blunt the cortisol stimulation response sent to the body (from the pituitary to adrenal glands) and therefore downregulate the cortisol feedback to the brain. As a result, with PTSD the body's cortisol feedback loops are damaged and dysregulated, which leads to increased inflammation because cortisol is a catabolic anti-inflammatory hormone.

A serious problem arising from exposure to chronic cortisol levels, as discussed in Chapter 4, is glucocorticoid resistance, a condition which alters the effectiveness of cortisol to regulate our body's inflammatory response to stress, injury and infections. Glucocorticoid receptor resistance has been reported in patients with chronic diseases such as asthma, depression, cancer, and cardiovascular diseases as well as susceptibility to diseases caused by respiratory cold or influenza viruses (such as Coronavirus type 229E)[21] or COVID-19[22]. Glucocorticoid resistance is the psychological, neuro-immune analog and precursor of insulin resistance, hence it wreaks havoc on the whole mind, psyche and even our immune response to viruses.

Our brain's response to prolonged and chronic stress has immense consequences for our mental and physical health. One can argue the neuroplastic changes in amygdala, hippocampus and frontal cortex, coupled with glucocorticoid and insulin resistance, form the common denominator and underlying mechanisms for the devastating 3D of diseases: dementia, depression and diabesity (diabetes + obesity). These diseases are confounded by our vulnerability to common psychological disorders discussed in chapter 4: The self-reinforcing (broken) feedback loops that condition us into perseverative fear, competition and reward seeking addictions.

But have these modern day chronic diseases ever been adaptive? According to the *pathogen-host defense* theory, depression must have been an adaptive mechanism in evolution in that it sequestered those infected with pathogens from the society to allow healing of the individual and prevent spread of disease. In my opinion, depression is an entirely metabolic process, a balancing mechanism and a feedback response to energy-consuming inflammations gone into overdrive. By reducing dopaminergic and energy consuming (cortisol-induced) activities, and reallocating metabolic resources, the human body can mount a stronger immune defense, thereby reducing the organism's risk of death[23].

As economic humans, we provide biohacking tools (medicine) to depressed or infected members of the society to ensure they promptly return to economic

Homeostasis/Equilibrium:
Small Loads and Imbalances

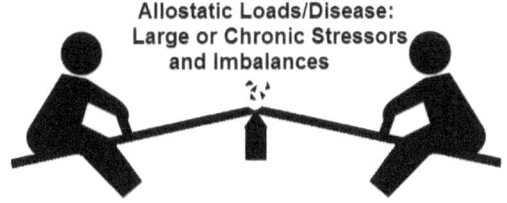

Allostatic Loads/Disease:
Large or Chronic Stressors
and Imbalances

productivity. This may help our economy but it prolongs both individual and societal stresses. Nature always needs *time* to allow for *adaptive healing*. But time, in the brain of a *Homo economicus*, equals economic productivity, income and opportunity cost. If the drug we use to mask or delay our depressive symptoms costs us less than our income opportunities, we often use the drug without any questions because it is economically efficient (adaptive). Drugs and biohacks help us convert depression (an adaptive mechanism) into *repression*. As the followers of Carl Jung know, *repression* often means serious trouble down the road.

Diseases and disorders are our brain's and body's adaptive tricks to handle a constant state of allostasis resulting from imbalances in our hormones and neurotransmitters. That is why chronic stressors in childhood are shown to result in epigenetic neurological changes in the brain. Children exposed to maltreatment develop smaller volume of the prefrontal cortex, greater activation of the hypothalamic-pituitary-adrenal axis (HPA) and sympathetic nervous reaction, and elevation in inflammation levels compared to non-maltreated children[24]. Through these disorders and diseases, our brains try to keep us alive, while metabolically managing a barrage of rewards and stressors never experienced by our early ancestors.

The Triggers of Self-Delusion in Human Brains

Our tools and (socioeconomic) rules which helped us establish scalable economies have also become our crutches. But when our brain and body are conditioned to ignore physical and natural feedback, we end up with a serious problem: Self-delusion!

Imagine looking at a window or screen that sometimes is transparent and allows you to see the (real) world behind it and sometimes, without any warning, projects a realistic *image* totally unrelated to what is behind the screen. You have probably seen these illusory projection screens in movies like *Mission Impossible- Ghost Protocol* (the scene inside the Kremlin where an optical camouflage screen is used as a subterfuge against Russian guards). If this type of screen is your only window to the outside world, it becomes nearly impossible to tell apart *real* world reflections (natural feedback) from *perceived and illusionary* reflections. Our brain is like that finicky deceptive window and screen that reflects its own perceived images. Illusions often become our self-delusions. In *The Believing Brain* author Michael Shermer provides numerous real-life examples to demonstrate how our brains, as construction engines for beliefs, reinforce them as truth in positive feedback loops.

Our modern brains are wired as conceptual, calculative, affectual and interpretive processors that are our windows to the world. Our eyes and noses are just providers of input to the brain. What we see or remember is the brain's output. In fact, new research[25] shows that episodic memories in humans are not merely a re-activation of stored experiences but a product of an intense *construction* process in which we manipulate our memories in such a way that we protect our positive self and mitigate negative memories that do not fit our self-image. That is how self-delusion works.

A self-delusional brain can override feedback loops from nature and our own body, and lead to allostatic diseases (already discussed) as well as biases, prejudices, self-deceptions, and psychological disorders such as schizophrenia, psychosis, hallucinations, and bipolar disorders.

There are many factors that could trigger self-delusional biases in human brains. The following are a sample of these factors, provided as a checklist for all of us to watch.

Conditioning: Fear, Reward, Competition

As discussed in chapter 4, human brains are uniquely prone to being conditioned by fear and anxiety (cortisol), competition (testosterone) and reward (dopamine) into self-reinforcing hormonal feedback loops. Imagine you know from reviewing historic reliable data that in your area there is only 1/1000 chance of being stung by a bee and when stung, a 1/1000 chance of showing serious allergic reactions. Your cortical brain calculates your odds of serious allergic reactions at 1/1000000 (one-in-a-million). But then your local TV news starts and continues for six months, showing local people who were hospitalized after being stung by a bee. This will condition many people with fear. Then for another six months, the TV station airs ads for a spray that repels bees from your backyard, conditioning you with a potential reward, and shows local neighbors have purchased it, conditioning you for competition. Because you are now conditioned to activate your trifecta of self-delusional conditioning circuits - fear, reward and competition - you will be very likely to ignore the rational objective 1/1000000 risk, be very afraid of the bees and buy the spray no matter the cost. You are not likely to change your mind even if I show you data confirming that the chemical spray actually poses a 1/100 (one-in-a-hundred) risk of cancer and infertility to you and your children. A judicious cortical brain would choose a one-millionth chance of allergic reactions over a one-hundredth (high) risk of cancer and infertility but when conditioned, our brain's self-reinforcing feedback loops are in the driver seat of our body and mind. In fact, the more invisible or physically intangible the threat is (such as an invisible enemy, a microscopic bacteria, a nano-scale virus, or a new disease or cancer), the easier it is for our brains to become self-delusional because we cannot physically assess the threat using our natural senses[xvii].

The problem with getting stuck in self-reinforcing feedback loops of anxiety, reward or competition is that they are exhausting and metabolically divert energy away from our energy-hungry frontal cortex and cripple it, so we cannot think rationally about data and nuances. Prudent analysis gives way to prejudgment (prejudices) which people mistakenly call judgment. In the words of Carl Jung: *"Thinking is difficult, therefore let the herd pronounce judgment!"*

Generally speaking, as humans, we are never far and immune from forming self-delusional thoughts and addictive traits because our concept-driven brain is not bound by nature's self-correcting feedback loops. We are *all* prone to delusional perceptions, obsessions, compulsions, schizophrenia and psychosis. The antidote to such vulnerabilities is also often behavioral, such as behavioral cognitive therapy (CBT) and other methods I will discuss in chapter 8.

Abstractions, Ideals and Idols

As discussed earlier, we think and communicate using a brain that imposes meanings on the outside world by defining conceptual categories that are often subjective and compartmentalized. If we hurt someone in a street fight, we end up in a prison but if we injure another human in a boxing arena or in a war zone, we may get applauded. A remarkable example of how human brains can be self-deluded by compartmentalization is the Christmas truce of World War I. On Christmas eve, 1914, in response to a plea by Pope Benedict XV, German and British soldiers stopped the fight and crossed into no-man's-land to mingle, play, sing carols, exchange food and souvenirs with each other. Two days later the war and bloodshed restarted. Future attempts at holiday ceasefires were not successful because soldiers were threatened with court marshals and harsh sentences. The story shows us how an abstract concept associated with peace

---

xvii Evolution did not give us a microscopic vision, maybe because the local diet and lifestyle of foragers helped them evolve a level of natural symbiosis with the local microbiome

and holiness (Christmas holiday) temporarily outpowered a nationalism concept (calling for war) in the brains of some soldiers. In the recent post-war era, the same holy concept is associated in the minds of many people with shopping.

Shared ideas and concepts can sometimes lead to superstitions. November 29, 2020 was the 9th day in the 9th month in the year 1399 in the Persian calendar, so it would be listed as 9/9/99. Iran's Birth Registry office was flooded with requests for this birth date for babies even born a few days earlier or later because the public believed 9/9/99 was a propitious date. Many parents had even planned the conception, months in advance, to end with a 9/9/99 birth date!

Belief in shared and sacred ideals can help humans strengthen their social bonds. The term neuroscientists and psychologists use now to describe a blessed higher state of being is *Eudaimonia*, the Greek word for a blessed euphoric state. Higher causes and ideals, whether religious, philosophical or political, can trigger our brains dopaminergic circuits and lead to euphoria, which is useful to fight depression, but also may contribute to psychosis and self-delusion.

One euphoric trigger of self-delusion is the *Quixotic* or *savior complex* which happens when humans see themselves in a parental role for others, and as the saviors of the planet or humanity. During the recent COVID-19 pandemic, some public health authorities assumed the role of savior of humanity and censored debate by penalizing and silencing independent scientists and even Nobel laureates. Other scientists with a savior complex urged the governments to conduct a deliberate challenge experiment in which healthy test subjects would be *deliberately* and unknowingly infected in randomized control trial studies. Obviously, these experiments violate basic principles of bioethics and The Nuremberg Code which bans human experimentation without the subject's informed consent and knowledge. People with a savior complex are often comfortable with forced mandates, *minority harm* and violation of basic human rights as long as it serves the majority interest.

Another example of a savior complex was provided in a report[26] which revealed that some of the vandals arrested in the 2020 Manhattan riots were children of multimillionaire families with vacation homes in Connecticut and memberships in yacht clubs and modeling agencies. These well-intentioned young wealthy rioters benefited from the eudaimonia associated with saving the poor against an unjust system, without realizing their lifestyles would probably not be possible without that very same system created by their parents.

Religious ideals could also lead to self-delusion as was already described in the case of the *Jerusalem Syndrome* which impacts some religious pilgrims.

Among the strongest sources of self-delusion in humans is affiliation with political parties or figures to the extent of idolization. Although not physical, political stimuli feel very real in human brains and can elicit real emotions and affects such as euphoria, anger, fear, anxiety and hatred, often to an irrational extent. Recently, the defunct Trump Plaza Hotel and Casino (once part of the bankrupt Trump Entertainment Resorts) was to be demolished (imploded) in Atlantic City, New Jersey. A group of wealthy political opponents of Donald Trump were willing to offer $175000 just for the pleasure of pressing the demolition button even though the building was no longer owned by Trump. $175000 is about three years of labor for an average American household. Regardless of how you feel about politics, could you imagine any other species in the natural world giving away the fruit of three years of labor for the pleasure of pressing a button to destroy a *symbol that used to be* associated with an opponent? On demolition day, many people paid $575 for a V.I.P. breakfast and a front-row seat in an oceanfront pavilion with a direct view of the loud and dirty implosion. Politics is about power and money so it is as real as wars to many humans now[xviii].

Our brains do a lot of categorization, abstraction and compartmentalization, all of which can lead to opinions, biases, prejudices or self-delusions. There are many humans who start a fight, with total strangers or even with friends and family members, in discord over sports, politics or religion. Humans also spend a good bit of their hard-earned income on abstract imaginary characters in movies, games or theme parks.

Metaphors

Although we know better, our brains often self-deludes us by interpreting metaphors *literally*. Cleanliness, for example, is a domain where the brain confuses the literal and metaphorical. In an interesting study,[27] researchers demonstrated how the brain has trouble distinguishing between being ethically dirty (immoral) and being physically filthy (in need of a bath). People who recalled an immoral act in their life were more likely to go for a package of antiseptic wipes after the experiment. Earlier in the chapter, I also shared the example of a 2010 political ad in which a New York gubernatorial primary candidate mailed out thousands of campaign ads impregnated with the smell of rotting garbage and photos of his rival, a scandal-tainted former Governor. The ad was obviously trying to associate the stench with unethical behavior of the opponent in a metaphorical way in the voters' minds.

Complexes and Unfulfilled Dreams

As an imaginative species, humans subconsciously carry their dreams and desires with them throughout their lives. Our powerful personal unconscious psyche is what Carl Jung calls a *complex*. There are countless stories of powerful and wealthy men and women who risk their entire fame or fortune for a fling or their life for a complex (like the tragic death of the model called Mexican Kardashian, discussed earlier). These irrational acts are driven by complexes, unfulfilled desires and goals. In some languages, the word used to describe

---

[xviii] For *Homo economicus*, the $575 would buy valuable networking with other politically aligned powerful folks which could pay off nicely down the road.

psychological complexes also means *"untied convoluted knots."* The *Sky's The Limit* attitude forms a feedback loop with our complexes and self-delusions.

Personal Gains, Wealth and Power

Humans self-delude when they personally gain from their own narration and interpretation of events. Personal gains are probably the reason scientists, doctors and officials who evaluated or prescribed medicines like Vioxx, Thalidomide, Fen-Phen, Baycol and Bextra became self-delusional and turned a blind eye to indicators of deadly flaws in these drugs that killed or harmed thousands of people.

In a 2017 article in The Atlantic[28] the author references research linking power to brain damage, literally:

"The historian Henry Adams was being metaphorical, not medical, when he described power as 'a sort of tumor that ends by killing the victim's sympathies.' But that's not far from where Dacher Keltner, a psychology professor at UC Berkeley, ended up after years of lab and field experiments. Subjects under the influence of power, he found in studies spanning two decades, acted as if they had suffered a traumatic brain injury—becoming more impulsive, less risk-aware, and, crucially, less adept at seeing things from other people's point of view. Sukhvinder Obhi, a neuroscientist at McMaster University, in Ontario, recently described something similar...When he put the heads of the powerful and the not-so-powerful under a transcranial-magnetic-stimulation machine, he found that power, in fact, impairs a specific neural process, 'mirroring,' that may be a cornerstone of empathy."

There are two classical studies that highlight the self-deceptive nature of human brains when it comes to power hierarchy. The first one is the 1961 experiment conducted by psychologist Stanley Milgram to measure the willingness of study participants to obey an authority figure who instructed them to perform acts conflicting with their personal conscience, such as administering electric shocks to help another human[xix] learn a topic. These fake electric shocks gradually increased to levels that would have been fatal had they been real. The results: 65% of participants, although some visibly uncomfortable, regardless of the level of education or income, administered the experiment's final massive 450-volt shock which could be deadly. They ignored their conscience in order to comply with the authority and to maintain their own authoritative power position over the (underdog) subject.

Another classical study, influenced by Milgram's 1961 electric shock experiment, shows that even a *perception* of power over another human is enough for our brains to self-deceive us. In the 1971 *Stanford prison experiment* volunteers were assigned to be either prison guards (perceived position of personal power and authority) or prisoners (underdogs) by the flip of a coin. Soon after the role-playing began, many students quickly embraced their assigned roles, with some guards enforcing authoritarian measures and ultimately subjecting some prisoners to psychological torture, while many prisoners passively accepted psychological abuse and, by the officers' requests, actively harassed other

---

[xix] In reality no one was tortured. Unknown to participants, an actor pretended to be shocked.

prisoners who tried to stop it. Several prisoners left mid-experiment, and the whole experiment was abandoned after six days because the role play had turned into serious behavior leading to incidents with real-life consequences. The 2010 movie *The Experiment* was inspired by the Stanford study.

Self-interest makes human brains vulnerable to self-delusional biases which blur the line between reality and imagination. In fact, personal gains are probably more powerful than religious or ideological convictions in eliciting prejudicial self-delusions in human brains. In the words of Upton Sinclair, the visionary writer and social activist: *"It is difficult to get a man to understand something when his salary depends upon his not understanding it."*

Herds, Mobs and Cults

We are not the only species that forms or joins herds in times of uncertainty. The human and bovine genomes are only 80% identical yet we both form herds because of evolutionary selection pressures such as predators. Herds offer adaptive benefits to central members of a herd, as opposed to outliers and loners. This is part of the *Selfish herd* theory of evolutionary biologist W. D. Hamilton whose *inclusive fitness theory* was discussed earlier.

Wild herds, flocks and colonies (such as bee hives) form distributed network brains. Swarm intelligence is advantageous because it reduces the survival cost to the individual. But *domesticated* herds, as opposed to wild herds, can be exploited and led to slaughter houses as in the case of cattle. Furthermore, unlike other domesticated (controlled) herd animals, humans are manipulated to form *cultish* herds, mobs and gangs that can become violent and tyrannical. As a result, individuals who join human herds, biased and self-deluded by conformity and sacred ideas, often act against their own rational interest or cognitive judgement. This is what happened to millions of Germans who joined the Nazi (National Socialist) cult.

The more mob-like a society becomes, the faster individual liberties and liberal democracies would disappear, leading to a dangerous vicious cycle. When herds become large enough, perhaps in times of uncertainty, anxiety or paranoia, self-delusion becomes subjectively *rational* as it serves a socially adaptive purpose for survival. *Cognitive conformity* is a strange self-delusional process in humans. Research shows[29] that irrational beliefs among humans can be easily induced through adaptive social cognitive processes that help with social integration. This is supported by the established role of mesolimbic dopamine in both delusions and social organization. In other words, when the reward (social integration and acceptance) of joining the Nazi party triggers release of dopamine in people's brains, their calculative rational brain (frontal cortex) would assign a benefit to conformity to Nazi doctrine and beliefs regardless of how objectively irrational or cruel they seem. This benefit is as real as the value assigned to caloric food for survival by our self-delusional brain. Psychologists call this phenomenon *cognitive dissonance* and the reason why humans who are caught lying, cheating and manipulating data often believe, *sincerely*, they acted rationally, honestly and innocently. As I mentioned earlier, assigning *intent* to a self-delusional brain is not a useful practice. We are all vulnerable to this conformist type of self-delusion, especially if we are well paid to conform to organizational or corporate guidelines and narratives.

Personally, I have been penalized or shunned several times in my life when I questioned the rationality of the conformist narratives. In one case, some 22 years ago, in one of the major scientific symposia attended by the leaders of the chemical industry, I raised a simple scientific question about the safety of a phthalate-based additive commonly used in plastics. The speaker and members of the expert panel did not know what to do with such an unprecedented question. For the rest of that symposium, some of the group-thinkers made me feel like an outcast. The same pressure for conformity exists in most other professional organizations, political parties, private corporations and government bureaucracies. Most censorship today is of the self-censorship, self-delusional type. Most deceivers today are sincere and true believers who can look us in the eye and misrepresent facts without flinching or blinking. Before we point a finger at another, we need to admit we are all prone to such self-preserving self-delusions.

A remarkable classical study in the conformity and self-delusional nature of the human brain was conducted by psychologist Solomon Asch[30]. Groups of eight male college students participated in a simple perceptual task to decide which of the three lines on the right has the same length as the line on the left. In reality, all but one of the participants were actors trying to fabricate consensus on a wrong answer 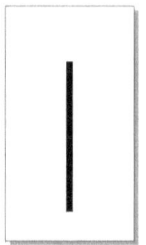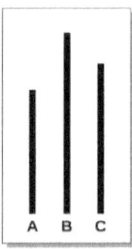 to bias the remaining participant (the real test subject). The true focus of the study was to evaluate how the remaining participant (always voting last) would react to the herd pressure towards an erroneous fabricated consensus. The study shows that 75% of participants gave at least one incorrect (self-delusional) answer to conform. In the control group, with no pressure to conform, the error rate on the critical stimuli was less than 1%. The good news is that 25% of the participants were independent objective thinkers who consistently defied the majority's fabricated opinion. These are probably the best humans to follow in times of mass hysteria, deception and paranoia.

In his opinion regarding the study results, Asch put it this way: *"That intelligent, well-meaning, young people are willing to call white black is a matter of concern."* Look around or check human history[31] to find countless examples of the Asch self-deception effect, from crusades, wars, witch hunts and mob lynching to modern day politically-driven mass hysteria (like McCarthyism), cancel culture and racism.

While early humans, like wild animals, may have benefited from herd formation and instincts, in modern times, we pay a cost for this behavior. One prominent historic example is ignoring the dangers of smoking. The earliest study that strongly linked smoking to lung carcinoma was released in 1950 by British epidemiologists Richard Doll and Austin Hill. Actually, five years earlier, in 1945, polymath and Nobel laureate Otto Warburg had already linked cancer to anaerobic glycolysis by mutagens triggered by smoking (as well as sugar, pesticides, chemicals, artificial food additives, electromagnetic radiation, and hypoxia). Yet, for the next half a century humans across the world smoked and suffered without knowing any of these early reports. Why? Because of the herd behavior led by self-deluded celebrities, scientists and even doctors who promoted the habit. Between 1950 and 1970, if you opened any magazines or turned on your TV, you would be hard pressed not to see a scientist or a doctor discussing the benefits of smoking certain brands. Our collective herd behavior as humans cost millions of lives.

Scientists are not immune from joining herds. A horrific case of conformity among scientists was the inhumane enforcement (mandating) of lobotomy (drilling holes in the skull to disconnect the frontal cortex) to treat psychological conditions and *moral insanity* in more than 100,000 men, women, children and homosexuals. Many died. Others like sisters of President John F. Kennedy and playwright Tennessee Williams were incapacitated for life. For years, critics were ignored or silenced by the powerful medical and science cults. We will discuss this more in chapter 7.

There are countless other cases in recent history in which herds and group-thinks proved harmful. A tragic example is the early days of polio vaccines that caused widespread injury in children due to defective vaccines and the bitter political and personal rivalry between two egotistical, politically powerful scientists, Jonas Salk and Albert Sabin, who advocated for two different vaccine technologies. Nobody was allowed to question the narrative. Many scientists joined the cult of Salk, and later Sabin. Dissenting scientists would be ignored, banished or silenced because the fear of polio, amplified on TV and by commercial entities, had created an atmosphere of paranoia in the society. Years passed before scientific idols could be questioned, mistakes and injuries were identified, and formulations were optimized. As usual, our children paid a heavy

price for the self-delusional cultish behavior of scientists, parents and authorities.

In a world of mobs and herds, being objective, independent and neutral becomes dangerous. Recently, a football referee in the town of La Jigua, Copán in Western Honduras was attacked by a lynching mob who did not like his neutrality. The referee was smart enough to carry a pistol while umpiring the match, which saved his life.

Specialization, Expertise

I have spent a good part of my life around diligent studious Ph.D., scientist and engineer types who are subject matter experts in their fields. They attend conferences and are nicely compensated for spending time on professional platforms to stay up-to-date on the fast-changing world of science and technology. But being a well-paid specialist does not make us immune from self-delusions. Judging from my own experience, staying up-to-date with my specialty was exhausting and left me unskilled to see the world outside my field of vision and expertise. I realized I was not the only one who was tunnel-visioned, self-delusional and biased among my fellow specialists.

Specialization today is akin to designing special keys for hard-to-decipher locks. Working a lifetime on deciphering a certain lock, the specialist sees everything through that keyhole. Specialists are our well-paid toolsmiths today. Yet real life problems are more like mazes, with many types of doors and locks.

Because specialization forces our brain to activate repeated routines, it is not uncommon to find some of the most celebrated names in science, technology, sports or business lacking *common street-smart sense* when it comes to their own personal lives, health, relations, marriage, children, politics, etc.

Free will is a controversial subject, but I believe exercising it requires what engineers call *degrees of freedom* in our mind. Being trained as a specialist often limits our brain's *degrees of freedom*. This is masterfully captured in Charlie Chaplin's 1936 movie *Modern Times*. It is a must-see if you have not watched it, particularly the factory floor scene.

Five years before *Modern Times*, Aldous Huxley, the savant author of *Brave New World*, also foreshadowed a future in which humans, conditioned to stay away from nature and books, become highly specialized and productive in their fields yet highly conformist, thoughtless and driven by self-indulgent pleasures. Their self-delusion, now their reality, was possible because they were sheltered from nature's feedback such as pain, stress and disease with *soma*, pills administered only by alpha humans who govern the society. In a world that highly rewards specialization, tunnel visions and self-delusions become common.

Rumination

Being unable to have an outside-in perspective makes us vulnerable to self-delusions. Rumination is a form of recursive self-focused thinking. Studies show[32] that rumination is involved in the etiology (root cause) and maintenance of major depressive and anxiety disorders. People with a ruminative thinking style often show a hyperactive limbic (amygdala) and prefrontal cortex, and an exaggerated prolonged cognitive response to severe life events. This can lead to depression or anxiety disorders when compounded by the subject's reactive temperament and current levels of depressive symptoms (including low self-esteem).

Buying into Logical Fallacies

When we have poor analytical skills and allow other humans to hoodwink us, we can become self-delusional. If not equipped to detect logical fallacies, we can reach foregone conclusions by short circuiting our deliberative frontal cortex. For example, a common logical fallacy is to reach *ad hominem* conclusions. We judge someone's arguments based on our impression of their personality. We prejudicially dismiss, in a wholesale fashion, any statement made by persons or groups blackmailed with unpopular labels such as anti-science, anti-religion, racist, unpatriotic, etc. I abhor labels and always try to listen directly to blackmailed people before I can analyze the entirety of their statements. But that takes time and people are busy so they trust sources and fact checkers who are often professional obfuscators paid to slant the story! There are many other logical fallacies which you can find in a good book on logic.[33]

Our Perceptions

As discussed earlier, perceptions become our brain's reality so if they delude our brain, they can impact our physiology and bodily functions. For example, our perception of time is more important than the real time in adjusting our metabolic rates. Research shows[34] blood sugar levels in people with type 2 diabetes follow *perceived* time rather than the actual time. Changes in blood glucose levels were measured in 46 participants with diabetes while they completed simple tasks during a 90-min period. Participants' perception of time was manipulated by having them refer to clocks that were either accurate or altered to run fast or slow. Blood glucose levels changed in accordance with how much time they *perceived* had passed (based on fake clocks) instead of how much time had actually passed. The results confirm the power of perceptions and brain constructs on biasing our brain and our body.

Another example of *perceptual bias* is our intolerance to noise (vs. signal). When our brains are fatigued and conditioned by fear, reward or competition, we cannot properly manage uncertainty, nuances, risk and anxiety outside our homogenous world and predictable routines. We will filter out or turn a blind eye to data or people outside our comfort zone as *noise*. We become perseverative and intolerant to feedback loops.

The same way noise intolerance and perseverative behavior are hallmarks of autism in individuals[xx], our societies are now collectively demonstrating metabolic fatigue and *social autism* by avoiding nuances and critical feedback.

The growth in the number of noise-intolerant people may also explain the phenomenon labeled as *The Cancel Culture*. The problem with silencing or ignoring different views is that it further aggravates the broken feedback loop that plagues perseverative people. This is how self-delusional echo chambers, group thinks, mobs and cults form.

Chapter Synopsis and References:

We tracked the parallel evolution of humans and their brains on a historic timeline to demonstrate the bifurcation of our evolutionary path from natural selection, mechanistically and not in a phylogenetic way. Our concept-driven brains are now in the driver seat of our mind-body homeostasis. The scalable economic rewards and the pursuit of quantity and speed, ushered in by the industrial age, were perfect conditioners for our brain's scale-free dopaminergic feedback loops leading to addictions and habituations. In a world built by *Homo economicus*, economic fitness replaces metabolic efficiency, symbiotic biodiversity and balance.

The Sky's the Limit attitude reflects our brain's simulated view of the world: A video gamer that can fly, run and fight countless opponents as long as s/he can respawn by *buying* lives and health. Our scalable rewards we collect (coins) allow us to *level up* indefinitely in the economic world. In the mind of *Homo economicus* the concept of money is the key to winning and surviving the economic game so it gradually supersedes old-world concepts such as courage, friendship, love, nationalism, and religion.

But boundless quantity and speed lead to imbalances, and therefore, disease (Dis Ease), discord and disorder. Our brains, having lost touch with natural self-balancing feedback loops, are prone to self-delusion, anxiety, depression and addictions.

[1] Ridley, Matt. "Why MRNA Vaccines Could Revolutionise Medicine." *The Spectator*, 19 Dec. 2020, https://www.spectator.co.uk/article/why-mrna-vaccines-could-revolutionise-medicine.
[2] Abigail E. Page et. al, "Reproductive trade-offs in extant hunter-gatherers suggest adaptive mechanism for the Neolithic expansion," *Proceedings of the National Academy of Sciences* Apr 2016, 113 (17) 4694-4699
[3] Clarke, Steve, et al. *Rethinking Moral Status*. Oxford University Press, 2021.
[4] Muthukrishna, Michael, et al. "The Cultural Brain Hypothesis: How Culture Drives Brain Expansion, Sociality, and Life History." *PLOS Computational Biology*, vol. 14, no. 11, 2018.
[5] Dehaene, Stanislas, and Elizabeth Brannon. *Space, Time and Number in The Brain: Searching for the Foundations of Mathematical Thought*. Academic, 2011.
[6] Krassioukov, Andrei, and Stacy Elliott. "Neural Control and Physiology of Sexual Function: Effect of Spinal Cord Injury." *Topics in Spinal Cord Injury Rehabilitation*, vol. 23, no. 1, 2017, pp. 1–10.
[7] Schnall, Simone, et al. "Disgust as Embodied Moral Judgment." *Personality and Social Psychology Bulletin*, vol. 34, no. 8, 2008, pp. 1096–1109.

[xx] New studies by Professor Martha R. Herbert have shown that autism spectrum disorder (ASD) is associated with difficulty in processing external "noise" and nuances due to high levels of endogenous neural noise in the brain of the autistic individual.

[8] Liberman, Peter, and David Pizarro. "All Politics Is Olfactory." *New York Times*, 23 Oct. 2010.

[9] Williams, Lawrence E., and John A. Bargh. "Experiencing Physical Warmth Promotes Interpersonal Warmth." *Science*, vol. 322, no. 5901, 2008, pp. 606–607.

[10] Levy, William B, and Victoria G. Calvert. "Communication Consumes 35 Times More Energy than Computation in the Human Cortex, but Both Costs Are Needed to Predict Synapse Number." *Proceedings of the National Academy of Sciences*, vol. 118, no. 18, 2021.

[11] Seeger, Pete, et al. "If I Had a Hammer: Songs of Hope & Struggle."

[12] Welniak–Kaminska, Marlena, et al. "Volumes of Brain Structures in Captive Wild-Type and Laboratory Rats: 7T Magnetic Resonance in VIVO Automatic Atlas-Based Study." *PLOS ONE*, vol. 14, no. 4, 2019.

[13] Pontzer, Herman, et al. "Metabolic Acceleration and the Evolution of Human Brain Size and Life History." *Nature*, vol. 533, no. 7603, 2016, pp. 390–392.

[14] Thomas, P.K. "The peripheral nervous system as a target for toxic substances," *Acta Neurol Scand Suppl.* 1984.

[15] Niimi, Naoko, et al. "Drug-Induced Demyelinating Neuropathies." *Advances in Experimental Medicine and Biology*, 2019, pp. 357–369.

[16] Wolman, Abel. "The Metabolism of Cities." *Scientific American*, vol. 213, no. 3, 1965, pp. 178–190.

[17] Jones, Luke. "Fluxopia: On Life in the Metabolic City." *Strelka Mag*, 27 Nov. 2020.

[18] Alciatore, Giacomo, et al. "Immune Challenges Increase Network Centrality in a Queenless Ant." *Proceedings of the Royal Society B: Biological Sciences*, vol. 288, no. 1958, 2021, p. 20211456.

[19] Harris, Julia J., et al. "Synaptic Energy Use and Supply." *Neuron*, vol. 75, no. 5, 2012, pp. 762–777.

[20] Sapolsky, Robert M. "Doubled-Edged Swords in the Biology of Conflict." *Frontiers in Psychology*, vol. 9, 20 Dec. 2018.

[21] Cohen, Sheldon, et al. "Psychological Stress and Susceptibility to the Common Cold." *New England Journal of Medicine*, vol. 325, no. 9, 1991, pp. 606–612.

[22] Guarnotta, Valentina, et al. "Glucocorticoid Excess and Covid-19 Disease." *Reviews in Endocrine and Metabolic Disorders*, 2020.

[23] Raison, Charles L, and Andrew H Miller. "Pathogen–Host Defense in the Evolution of Depression: Insights into Epidemiology, Genetics, Bioregional Differences and Female Preponderance." *Neuropsychopharmacology*, vol. 42, no. 1, 2016, pp. 5–27.

[24] Danese, Andrea, and Bruce S. McEwen. "Adverse Childhood Experiences, Allostasis, Allostatic Load, and Age-Related Disease." *Physiology & Behavior*, vol. 106, no. 1, 2012, pp. 29–39.

[25] Dings, Roy, and Albert Newen. "Constructing the Past: The Relevance of the Narrative Self in Modulating Episodic Memory." *Review of Philosophy and Psychology*, 2021.

[26] Fonrouge, Gabrielle. "Inside the Privileged Lives of Protesters Busted for Rioting in Manhattan." *New York Post*, New York Post, 10 Sept. 2020.

[27] Zhong, Chen-Bo, and Katie Liljenquist. "Washing Away Your Sins: Threatened Morality and Physical Cleansing." *Science*, vol. 313, no. 5792, 2006, pp. 1451–1452.

[28] Useem, Jerry. "Power Causes Brain Damage." *The Atlantic*, Atlantic Media Company, 23 June 2017, https://www.theatlantic.com/magazine/archive/2017/07/power-causes-brain-damage/528711/.

[29] Bell, Vaughan, et al. "Derationalizing Delusions." *Clinical Psychological Science*, vol. 9, no. 1, 2020, pp. 24–37.

[30] Asch, Solomon E. "Studies of Independence and Conformity: I. A Minority of One against a Unanimous Majority." *Psychological Monographs: General and Applied*, vol. 70, no. 9, 1956, pp. 1–70.

[31] Bagus, Philipp, et al. "Covid-19 and the Political Economy of Mass Hysteria." *International Journal of Environmental Research and Public Health*, vol. 18, no. 4, 2021, p. 1376.

[32] McLaughlin, Katie A., and Susan Nolen-Hoeksema. "Rumination as a Transdiagnostic Factor in Depression and Anxiety." *Behaviour Research and Therapy*, vol. 49, no. 3, 2011, pp. 186–193.

[33] "Logical Fallacies." *Logical Fallacies - List of Logical Fallacies with Examples*, https://www.logicalfallacies.org/.

[34] Park, Chanmo, et al. "Blood Sugar Level Follows Perceived Time Rather than Actual Time in People with Type 2 Diabetes." *Proceedings of the National Academy of Sciences*, vol. 113, no. 29, 2016, pp. 8168–8170.

Chapter 6: Masks, Crutches and Daggers

"All the world's a stage, And all the men and women merely players"
- William Shakespeare

Epigraph: As we metamorphose into *Homo economicus*, a metabolically inefficient species handicapped in processing self-correcting natural feedback loops, we rely on masks, crutches and daggers as our new adaptive tools. These may be beneficial in the short-term but further alienate us from nature's balancing and self-correcting feedback. I share how reliance on masks and crutches did not evolutionarily bode well for crafty intelligent species like cephalopods.

The Adaptation of a Denatured Species

Humans are uniquely equipped with an imaginative, calculative, creative brain which has evolved to define stresses and rewards for our body and give us the luxury of ignoring natural feedback loops. With our concept-driven brains being in the driver seat of our bodies, we are a denatured species, one which psychiatrist George Engel calls a biopsychosocial species. As we already learned, our brains are easily, self-delusionally and addictively conditioned by fear, reward and competition. At the same time, we use tools (crutches) and ploys (daggers) that help us ignore and mask any biopsychosocial feedback signal. In the world we have built, as I exemplified using the stories of my own life as well as those of Howard Stern and Roseanne Barr, anyone with any level of trauma and imbalance can move up the socioeconomic hierarchy by acquiring better masks - tools to externally cover up our imbalances - , crutches - tools to internally mask our imbalance and trauma, and daggers - ploys to eliminate rivals.

Masks

Masks make humans a dangerous species. Ghislaine Maxwell, who is now on trial as an accomplice of the notorious underage sex slave trafficker Jeffrey Epstein, was a guest of honor at The New York Society For The Prevention Of Cruelty To Children's 2013 Spring Luncheon at The Pierre Hotel in New York City[1].

Game Theory, which is a series of mathematical algorithms to model interaction among social organisms to predict social outcomes, is not easily applicable in its classical form to humans because of the extent to which we play mind games. Humans can not only fool others by masking their intentions and flaws, they can easily self-delude themselves to become believers in the mask they wear. Our masks often become part of our persona and given that humans do not rely on natural biological balancing feedback loops, we may not even recognize how many layers of masks we are wearing. William Shakespeare was wise by noticing: *"All the world's a stage, and all the men and women merely players."* And the more we rely on social acceptance instead of natural feedback, the more adept we become in layering up masks, which makes actors and politicians particularly vulnerable.

The renowned psychoanalyst and psychiatrist Carl Jung uses the word *persona* as one of the main archetypes in our collective human unconscious. In other words, he believes acting to socially fit in has become an inseparable part of human psychological evolution. The word *persona* is derived from the Latin word for a mask worn by actors. In Jungian psychology the persona archetype enables one to portray a character, which is not necessarily his or her own, with the intention of presenting a favorable impression for social acceptance. Jung believes masks (persona) are adaptive but wearing too many of them becomes maladaptive, as described later in this chapter.

Because of our self-delusional nature and our love of mind games, it is practically impossible to separate people from the masks they have worn for years. For most of us that inner child is long gone and deeply buried[i]. As adults, we have little or no concept of ourselves as beings distinct from what society expects of us. According to Jung, persona is a mask that feigns individuality, but is indeed a reflection of the "collective" psyche. The opposite of a Jungian world of inflated egos and persona is a utopian concept called noosphere first proposed by biogeochemist Vladimir Vernadsky, and philosopher and Jesuit priest Pierre Teilhard de Chardinrecent. A noosphere is a thought sphere that encompasses our collective global human consciousness, and is the evolutionary equivalent of a biosphere as a human-contributed layer on the planet.

I will describe later how the masks we wear are among the main causes of human suffering. So in what follows, I outline some forms and manifestations of masks commonly in use among modern humans.

Civilized Speech and Behavior

Nature is about real-time instant, and sometimes brutal, feedback. Early human tribes exchanged real time feedback, sometimes positively, as in selfless hospitality, and sometimes violently, as in honor killings or revenge. If you lived in that type of world, you would quickly learn everyone's red lines and avoid

---

[i] In *The Truth is a Lie*, author Harry Petsanis describes how our masks, now part of our identity, have emotionally tied us up in knots so we are afraid to reveal ourselves to people and admit that at heart we are all self-serving beings.

minefields. You would also learn what motivates people rather easily as is the case with primates. Most primate studies use motivators such as food to encourage learning or modify behaviors. In what we call the uncivilized world, organisms are easy to figure out. Happy, angry or fearful faces are all real, and motivations are sincere, heartfelt and easy to detect and predict (or manipulate by a modern human).

Without language, natural feedback is straightforward and difficult to hide. That is why unlike humans, animals cannot like or even easily mask their signals and feedback. As mentioned earlier, a dog's anal glands release distinct pheromones that reveal the dog's health, age, sex and other affinities and vulnerabilities. When threatened, dogs can only cover up their vulnerabilities by tucking in their tails to block release of the revealing anal pheromones.

Now let's fast forward to civilized human societies formed around cities housing people with advanced language skills who let out their anger, fear and pleasure not in real-time simple words or reactions, as in primitive societies, but cloaked in well-calculated, well-measured demeanor and indirect words. We call these soft social skills. One of the best practitioners of soft skills in modern history is Winston Churchill. As compared to his brutal arch nemesis, Hitler, who used aggressive ideological language, Churchill was a master of soft illusionary, sometimes even romantic, speech. For example, in the worst days of the war, he moralized the destitute British citizens being bombed every day by promising them a victory in *"God's Good Time."* Not many other orators could come up with two amazingly uplifting words "God" and "Good" to convey a message that basically broke the bad news to citizens that meant "You need to resist and endure misery for a long time." If you cautiously read Churchill's speeches, you will find how amazingly light and playful his language was. He was a smoker, a drinker and of a sickly temperament yet an effective war-time communicator, unrivaled in masking bad news in uplifting speeches to the citizens.

The civilized world is a world of mind games and speech craft. That is why it takes some 25 years to train a child in deciphering society's signals. Without soft skills, humans often end up dead or in prisons in civil societies. A lawyer and a poorly educated man can receive grossly different sentences for committing the same misdemeanor because the former knows the proper jargon and speech that would mask the severity of the misdemeanor in a civil society. Our civilized speech and behavior is a mask we need to wear to survive in a civil society.

Shakespeare has compared (in King Lear, Act I, Scene IV) well-spoken (educated) folks who are vain, hypocritical and spineless to court jesters and fools serving the king: *""Fools had ne'er less wit in a year, For wise men are grown foppish; They know not how their wits to wear, Their manners are so apish!"*

Polished or Rebellious Appearance and Decorum

In modern civil societies, propriety applies not only to our words but also to our looks and decorum. Polished shoes, neatly coiffed hair, looks and demeanor often gain other human's trust more than disheveled looks, dirty attire, and crude behavior. To some people like me, a super polished appearance and decorum signal discomfort or insincerity, but I am probably in the minority. Smooth-looking or smooth-talking politicians like John Edwards, who was extremely polished and well-coiffed (so he lost my trust), are often popular with voters. Edwards was almost elected Vice President of the U.S. in the 2004 election. His popularity only waned after it was revealed that while his wife, Elizabeth, was dying of cancer he had fathered a child in an extramarital affair.

Several research papers confirm the human's positive bias towards beauty and polished appearance. This is probably why the young unkempt Albert Einstein had a hard time finding a teaching job. There are probably many other capable and accomplished humans like Einstein who are undervalued by the society because of their disheveled appearance. And most of us know looks do not mean brains or integrity yet we like to fool ourselves because remember we are a self-delusional species.

The opposite of a polished appearance, is the rebel stereotype adopted by youngsters and people who in their personal unconscious harbor some resentment they project into their socially rebellious persona. If not followed by faithful acts and meaningful sacrifices in personal life, rebellion can become virtue signaling and also a prosocial mask.

Prosocial and Virtue Signaling

Virtue signals (public expressions to show off one's virtues) are now a widespread form of masks among humans. The world is now full of justice fighters and bleeding heart advocates for peace, minority rights, education, health, veterans, environmental protection and other socially popular causes. Yet generally speaking, we are not doing well in metrics related to education, health, social justice, environmental protection, or peace[ii]. That is because real virtues are often more costly and less socially rewarding than virtue signals. One of my popular adages used by Quakers who are known for peace activism is "Let Your Life Speak." Another one is by the 17th century writer François de La Rochefoucauld: "*Virtues are lost in self-interest as rivers are lost in the sea.*" Yet if a virtue is signaled and repeated enough, many humans will start following the signalers. Just think about social media. The top ten most popular Twitter accounts now reach a combined 400 million accounts on the planet, unmatched historically by any prophets, sages or philosophers advocating for peace, contentment, justice and health. Our top influencers are all for equity, justice and humanity yet we, as their followers, are still plagued with divisions, diseases, greed and injustice to each other. Why? Because we know virtues spoken by our favorite social media gurus are mostly signals and masks yet we follow them because in Jungian terms, they represent some collective unconscious among us,

---

[ii] Annual medical bills budget in the U.S. now exceeds $1.4 trillion (excluding COVID-19 vaccine budget) and the annual war/defense budget now exceeds $700 billion.

maybe a sense of guilt, need for inclusion or acceptance in society, and aspirations to be virtuous (without paying the costs).

It is not easy to go against mobs and to question powerful virtue saints in popular culture. Today's independent comedians, satirists and podcasters like Joe Rogan, Tim Dillon and Jimmy Dore, who have become social commentators serve a much-needed purpose to expose and challenge our hypocritical virtue signaling tendencies. I used to like George Carlin, whose sharp tongue challenged the powerful and their mainstream narratives. There are still a few independent social commentators and satirists who are equal opportunity challengers of virtue saints. I may not agree with all their viewpoints but I am glad America still has these critical outlier voices not jumping on the mobs' virtue carnivals and bandwagons, or repeating the lucrative partisan political narratives. These independents are mirrors to our mask-wearing, self-delusional, and mob-forming species. I hope they don't get corrupted by money or silenced by power.

Crutches

Many of our tools have become crutches of convenience and not adaptation. An adaptive tool would utilize, enhance and strengthen our natural faculties and abilities. A crutch on the other hand, would shunt natural feedback loops by weakening or bypassing our natural evolution. Unfortunately many electronic gadgets, drugs and pharmaceuticals fit in this category because they block natural feedback loops available to us. Let's look at alcohol as an example. Despite recent studies that show its harmful physical effects even in small doses[2], we often misuse alcohol as a crutch and develop an unhealthy dependence on it to facilitate our prosocial and stress management skills.

Alcohol mediates a change in our behavior by shunting and distorting the natural and balancing feedback loops (psychological and biological) inside and outside us. In nature, a healthy colony relies on biological and social reactions and self-adjusting (self-correcting) feedback loops to balance itself. Anything that dilutes or distorts these real and natural feedbacks more or less fits the definition of a crutch. It can be alcohol or a mobile phone that has replaced our face-to-face feedback loops.

Crutches also artificially alter our natural neurological and physiological baselines and homeostatic set points. Take our body temperature as an example. In an average human, cells function optimally between 98 °F (37 °C) and 100 °F (37.8 °C), which is the homeostatic safety window for our temperature baseline and set point. When we are physically infected by pathogens, the energy released by our body's defense mechanism may cause a temporary elevation of our temperature beyond the safety window. We call this a fever. When we are depressed, our temperature baseline (set point) also increases but by only about 0.5 °F (0.3 °C). Our whole body's homeostasis works through a complex system of evolutionarily interdependent hormones and neurotransmitters to ensure we return to our baseline's healthy temperatures by addressing the underlying issues causing the elevated temperatures. Now if we regularly use a slew of drugs and interventions to speed up the temperature recovery, we may be masking the underlying root causes of our imbalance. This is one of the reasons why crutches often lead to masking deep problems.

What Turns a Tool into a Crutch?

1) A tool temporarily assists or nourishes our own physical and psychological abilities but a crutch makes us dependent on it for the long term so it often ends up numbing and weakening our own faculties. Crutches may make us feel good when we face problems but they usually do not make us stronger the next time we face the problem. In a way, a crutch is yet another mask because it covers up and masks our shortcomings. Imagine a heavy-set person wearing a corset to look thinner in the midsection. The corset becomes a mask shielding the person from the outside world's feedback. But the real problem with a crutch is not blocking external feedback but making us forget about our internal psychobiological feedback, for example the visceral fat being part of a feedback loop signaling excess cortisol, insulin resistance, metabolic disease and a suppressed immune system.

2) A crutch bypasses natural balancing feedback loops so it is often used excessively and becomes addictive. For example, in the early days of X-ray invention, it was used excessively as a crutch. Emil Grubbe was the flamboyant and ambitious doctor who first used X-ray to detect and treat cancer, but the overuse of the technique on himself and others (as a crutch), led to his own death from cancer despite undergoing 90 surgeries. Many scientists who exuberantly worked on radioactive materials in the early years also had tragic outcomes.

3) A crutch often helps us kick costs down the road so it offers attractive short term rewards at the expense of long term costs. As a result, crutches often beget more crutches. As explained in chapter 4, when we offer our brain easy fast rewards (low-hanging fruit) and short-term pain avoidance we are conditioning our brain's dopaminergic circuit into a self-reinforcing (addictive) cycle. This is why even mild narcotics could act as a gateway to stronger ones that are habit forming and addictive. Many legal drugs can also lead to dependence and act as gateways to additional drugs to curb the harms induced by the early crutch down the road.

Let's look at some historic tools that have turned into crutches. Shoes are a helpful tool developed by humans. We have even invented shoes (horseshoes) for our horses. Yet, over time, the soles and heels on our originally minimalist shoes have become thicker despite new studies demonstrating that *minimalist* shoes and sandals with thin flexible soles that mimic walking barefoot offer us anatomical, metabolic and evolutionary benefits. This is exactly what Ted McDonald, nicknamed Barefoot Ted, a spiritual naturalist athlete with many 21K ultra-marathons in his record, has discovered and shared in *Born to Run*,

Christopher McDougall's best-selling book about the evolution and science of endurance running.

Raised in California's outdoorsy *Surf and Skate* nature-embracing culture of the 1970s, before the age of fancy video games and cell phones, McDonald discovered first hand that being barefoot or in sandals, on a sandy beach, on a surfboard or a skateboard, or while running, is one of life's simplest and most intense natural pleasures. The natural grip of barefoot or minimalist shoes not only *grounds* us (electrically and immunologically) but also allows our body to best balance itself on all types of terrain.

The book also features the story of the indigenous people of the Copper Canyons in northwestern Mexico called *Rarámuri*, which translates in English to *those who run fast*. The Rarámuri are considered some of the best long-distance runners in the world because they use a midfoot strike (MFS) or forefront strike (FFS) technique for remaining light on their feet. This is possible because while running, they wear minimalist sandals or no shoes at all.

Unfortunately, many modern shoe designs are often anything but minimalist and have turned into crutches that impose both metabolic and anatomic costs on our body, especially the joints in our knees.

Another modern crutch is the medicines we overuse and abuse. We have increased our drug prescriptions in the U.S. from $22 per person each year in 1970 to about $1200 per person in 2020. There were only three main childhood vaccines in 1970: Diphtheria, tetanus and pertussis. Currently, 16-18 vaccines are administered, some requiring multiple doses, at different ages. Is it because we are facing many more infectious and viral diseases and immune compromised people today than 50 years ago? Probably not! Public hygiene, nutritional knowledge and medical standards are a lot better today than half a century ago. Additionally, many more people smoked back then than today, which made them more vulnerable to diseases. The reason for our rampant use of drugs and vaccines is our abusing them as crutches instead of addressing the underlying lifestyle and dietary conditions for our diseases. By definition, Crutches *always* beget more crutches.

Another group of widely used, albeit helpful, tools that function as crutches to many of us are glasses and contact lenses. Without them many of us cannot see clearly. Evolutionary biologists believe myopia is a relatively recent disorder in our evolutionary timeline coinciding with our move away from open meadows and farms into urban areas, close quarters, small offices and in front of computer monitors, TV screens and mobile gadgets. Myopia is evolutionarily maladaptive so it cannot be a common disorder in other species. Yet we treat the disorder with crutches that allow us, collectively, to ignore other crutches which caused it in the first place.

Behaviors can also turn into crutches. For example, excessive or out of control eating, shopping, hoarding, or even exercise could become crutches if they are used regularly to curb depression, anxiety or mood disorders.

Daggers

Masking our vulnerabilities and using crutches weakens us physically and metabolically but strengthens our brains skills in mind games and role playing. Weaker physical traits and stronger deceptive skills mean we carry daggers with us to survive in the human jungle. Daggers are our hidden and harmful resources or traits that give us a competitive advantage over another human. The financially powerful rely on daggers such as blackmailing (secret "dirt" on someone), PR (public relations) firms, media pundits and mercenary journalists, lawyers and even hitmen and fixers. Ordinary people rely on simpler daggers such as lying and bad-mouthing someone with their boss, family or friends. Social media allows the use of daggers on a mass scale. There is a lot of doxing and trolling out there.

Living as a human is not that easy if you think about it. We are not as blessed as we think we are. We spend a good part of our life and energy trying to figure out who is wearing what mask or hiding which dagger, what mask to wear and which crutch or dagger to use. Meanwhile, we are all prone to self-delusional addictive habits and broken feedback loops. This means for a good part of our lives, we experience a stressed state, a sympathetic nervous system, and the imbalance and diseases that come with the cortisol overload.

Will Masks, Crutches and Daggers Lead to Our Extinction?

Masks, crutches and daggers make us stressed, physically weak (dependent, metabolically inefficient and maladaptive), and intellectually deceptive and self-delusional. A colony of masked, crutched stabbers may not last for too long. The following are some of the other problems associated with evolving as a species independent of natural feedback loops and dependent on masks, crutches and daggers.

Inflated Tyrannical Egos plus Inferiority Complex

Carl Jung believed the overuse of persona (masks) is maladaptive because it would lead to our ego (the self-preserving gatekeeper of our conscious individual character) identifying with the persona, inflating it as a self-important role, imposing it on other humans and turning us into little tyrants. Self-righteous tyrants with inflated egos are often victims of their own persona, and live with feelings of inferiority and hypocrisy because they are alienated from their true nature. The solution to an inflated ego is to *deflate* it. Apart from experiencing hard setbacks in life (such as hitting rock bottom or near-death experiences), spiritual or meditative practices (many involve deep breathing and relaxation), or use of certain psychedelics (that would dissociate and decouple our calculative ego-driven cortical brain from our limbic and subconscious brain), it is not easy to break free from an inflated ego and from the masks and veneers with which we have identified for years.

According to Jung, when our ego reaches a state of balance with our evolutionary (animal) instincts (*shadow* in Jungian parlor) buried and suppressed deep in our unconscious, we may not need to wear many masks anymore. This may be why self-deprecating truth about our deep instincts, follies and desires (such as those offered by comedians) can help us tame our ego and bring to the surface the buried unconscious which we usually suppress.

Maladaptive Individualism: The Story of Cephalopods

Humans can learn a lesson or two from cephalopods. Like *Homo economicus*, cephalopods are all brains and disguise (masks). When it comes to bluffing and covering up, cephalopods - literally meaning brain in feet (arms) - are masters of (underwater) disguise and escape. After their ancestors lost their protective shells and became physically vulnerable (as humans who have lost hunting and foraging skills) cephalopods evolved problem solving brains, distributed in their limbs, which helped them adapt to predatory selection pressures. A veined octopus, for instance, demonstrates sophisticated tool use behavior by pulling two halves of a coconut shell together to hide inside. Once done, the animal stacks the shells together like nesting bowls and hauls them away because they may prove useful in the future again. So like humans, they plan and use tools and depend on them as crutches.

Another Cephalopod with a sophisticated predator-avoidance adaptation is the Cuttlefish, which is known as the master of disguise, because it can *bluff* predators through forming eyespots on its body which make it appear like a giant fish. They only use this trick against predators that rely on vision to find prey. Around predators that depend on their sense of smell, the cuttlefish is smart enough just to flee. If that does not work, Cuttlefish can easily shape-shift and change colors to look like squid, crabs, algae or even rocks! Cuttlefish even demonstrate intelligent, strategic, and delayed gratification decision-making based on previous experiences[3].

Despite their complex predator-avoidance intelligence, cephalopods live short lives as compared to many less intelligent species. This is perhaps because they mostly live solitary (individualistic) lives. Without any long-lasting social bonds, which confer on eusocial species additional safety nets against predatory conditions, cephalopods bear the metabolic costs of stressful camouflaging and deception alone.

If we are to learn anything from one of the most deceptive shape shifting species in nature is that relying on deceptive masks and crutches, will lead to a short-lived solitary stressed existence. We, humans, may be well on our way down this path.

Governments in several countries (UK, Japan, Germany and Canada) are establishing Ministries of Loneliness to address the crisis of individualism. But we cannot continue using masks, crutches and daggers, and expect *not to be* lonely. We will become a fragile species cocooned in protective masks and shells when we escape feedback and have others bear the costs of our actions.

Rampant violence, suicide, mental and physical abuse and disorders, all result from broken feedback loops.

Crutches Will Make Us Obsolete

We may think that advanced tools and crutches will extend our natural abilities but many recent examples show that they will make us obsolete. An example which I personally witnessed some 25 years ago in my corporate job was the rollout of enterprise customer and sales management software that would allow our sales and technical teams to handle and process data efficiently. We moved from taking mental and paper notes on our personal relations with our colleagues, vendors and customers to entering detailed quantitative notes in large central databases. Once the database was complete, it replaced (i.e. laid off) many sales and technical managers whose network and customer knowledge were no longer a precious asset or a well-kept secret.

Human obsolescence by crutches we have developed continues on a large scale. Computer programmers have written automation programs that have helped manufacturers and employers replace manual laborers. Now programmers are now writing code-generating AI software using platforms like OpenAI that may replace the programmers themselves by allowing end users to generate customized code on their own.

On a physical level, depending on crutches makes us physically unfit. Many of us are suffering from anatomical atrophies when compared to our ancestors. We already discussed the roots of myopia. Even walking and climbing, the two basic locomotive functions of primates and early hominins that kept them metabolically fit and balanced, are not easy tasks for many humans today because we depend (as a crutch) on indoor gadgets, supermarket food, sedentary lifestyles and jobs. As a result, as people's legs are atrophied, they develop visceral fat and metabolic disorders such as obesity and diabetes.

The Exhausting Psychodynamics of Masks

Maintaining masks are exhausting and metabolically expensive even for masters of disguise. In cephalopods, the large number of chromatophore cells which are responsible for their color change during camouflage carries a metabolic cost that impacts on the organism's energy budget. Changing colors is metabolically costly for humans too. Carl Jung even uses the laws of thermodynamics (discussed in chapter 2) to explain the root of our internal struggles between our individual conscious psyche and our collective unconscious psyche (masks). According to Jung, like all other systems seeking dynamic energy equilibrium, every psychic development in humans, if not close to a state of balance, will have behavioral consequences in the person's life. For example, the person with a social *strong man* persona who looks and acts tough, in his private life or when older or in crisis, may exhibit *child-like* behavior where his own feelings are concerned.

As discussed in chapter 4, modern neuroscience does confirm that our nervous system and our brain's neuroplasticity are indeed governed by metabolic homeostasis and energy equilibria. This validates the broad precepts of the science of psychodynamics, first proposed by psychiatrists like Jung and Freud some 120 years ago, that combines thermodynamics and neurology to describe psychological and transient functions of the mind.

So psychodynamically, wearing too many masks for too long is a tolling process. As Jung puts it *"A man cannot get rid of himself in favor of an artificial personality without punishment."*

Masks are also costly because we all spend a lot of our energy to see behind them the true character and intentions of the wearer. A good part of our social capital and energy is spent on catching conmen or becoming one! Our legal system, often expensive and overloaded, spends a lot of resources on proving or disproving intentions, in other words, removing masks or figuring out the conscious or unconscious roots of the perpetrators' behavior and psyche, something that even renowned psychoanalysts like Jung could not easily figure out. Many countries as well as the U.S. have adopted complex judicial and sentencing systems by classifying murders into two degrees (first- and second-degree), and manslaughter into voluntary and involuntary, all based on figuring out the exact type and timing of the perpetrator's *intent*. Remember, millions of hours in human effort and billions of dollars in our resources are being spent to unmask the intentions of a species that easily self-deludes and changes personas.

According to Jung's theory, our masks cover up our complexes, the buried mini-personalities that are unconsciously but autonomously driving us. Among the strongest complexes that I have observed or experienced are money, fame, power (influence) and sexual complexes. Because these complexes are unconscious and often unresolved, countless humans are driven by them without awareness. Neurologically speaking, complexes start and feed off of those self-delusional and self-reinforcing addictive reward (dopamine), fear (cortisol) and competition (testosterone) conditioning circuits discussed in chapters 4 and 5.

Like other unconscious drivers of our psyche, complexes can be metabolically (psychodynamically) exhausting. Without resolution, complexes continue to exert unconscious, maladaptive influence on our thoughts, feelings, and behavior and keep us from achieving psychological integration with our conscious self. Jung uses the analogy of worms which always rot the core. He believed that with our masks and complexes, we are often waging wars on *ourselves*.

Masks, Crutches and Daggers Beget More of the Same

Our brains are easily conditioned into cycles of addictive reward seeking, risk avoidance and endless competition, and our masks, crutches and daggers allow us to maintain these cycles. Masks help us reap easier rewards, crutches help us deflect or ignore risks, and daggers allow us to stay ahead of our cut-throat competition.

A stark example of how crutches beget crutches can be found in the way we abuse medicines that mitigate one disorder but induce others. For example, the following drugs can all upset our natural metabolism and raise our homeostatic blood glucose (sugar) levels: statins (cholesterol), steroids (rheumatoid arthritis, lupus, asthma, allergies), drugs that treat anxiety, ADHD, depression and other psychological disorders; birth control pills; blood pressure medication such as beta blockers and diuretics; decongestants containing pseudoephedrine and cough syrups. So prolonged use of a medicine as a crutch to treat symptoms of one disorder may end up inducing other major disorders (pre-diabetes or diabetes in this case), which require new crutches.

The small medicine cabinets that popped up in homes around 70 years ago are now filled with all kinds of synthetic compounds. Some people have cabinets and cupboards full of medications. My late father had a sophisticated pill dispenser, given to him for free by the pharmacy, to sort out the 8-10 pills a day he was taking during the last years of his life. I remember when he was in his 30s he occasionally took an aspirin for his headaches or a sleep aid. Like many others, he progressively depended on more medicines as he aged beyond 50. In his 60s, the pills he was prescribed allowed him the freedom to eat the kind of salty, sugary, fatty food that he liked. But in his 70s he started to see the impact of the pills and his diet on his kidneys and regretted some of his earlier decisions.

According to the Kaiser Family Foundation the number of retail prescription drugs filled at U.S. pharmacies in 2019 was about 12 per capita, as compared to 5.5 in 1972. Average cost of a prescription spiked from about $4 in 1972 to nearly $100 in 2019. This is evidenced by the progressive increase in the use of medicine in different age groups. Starting at age 18, the pharmaceutical expenditure per person almost doubles up every 15 years so the 50-64 age group spends twice per person on medicine than the 35-49 age group. And on the extreme end of the drug use spectrum are those who become dependent on the most potent crutch of all drugs, opioids. That is usually the *terminal* crutch. Tragically, in 2020 an estimated 92000 Americans died overdosing on opioids, both legally prescribed types and illegally sold on the streets of America (synthetics like fentanyl).

Groups such as Robert F. Kennedy's Children's Health Defense organization blame the pharmaceutical companies and regulatory agencies for pushing medicines and vaccines on the population. But I believe like all economic transactions, the *demand side's* pull (by consumers) plays a major role in how the market functions. Like my parents and siblings, a large number of people now *depend* on medical interventions as easy crutches to their imbalances (disease) that help them avoid difficult changes to their diet and lifestyle.

Electronic gadgets are also potent crutches. Among my family and friends, many of them feel obliged to have the newest smartphones, cars, TVs and laptops regardless of the cost. I know many people who cannot even go for a walk without their smartwatch that counts their 10,000 steps, the exact number they need to take every day to stay fit. These people do not realize the crutch they are using to stay fit is extremely active in connecting to Bluetooth, Wi-Fi, and cellular networks and exposes them to low-powered but potentially harmful radiofrequency (RF)[4].

Masks Beget Labels and Mobs

Our masks take us away from acting upon our individual conscience and make it easier for us to fit into stereotypes and labels. Today, billions of humans brand themselves or each other with labels like leftist, rightist, socialist, racist, anti-vaxxer, anti-Semite, antichrist, etc. I believe labels paint another human with a broad brush and ignore the fact that humans have independent, conscientious brains and souls. Labels turn us into effigies that can be easily hated, suppressed and divided. I believe masks and labels play major roles in stoking our social disorders.

Carl Jung is known to say: *"Thinking is difficult, therefore let the herd pronounce judgment."* Earlier I shared why according to Hamilton's *Selfish Herd* theory, sometimes in nature there are evolutionary advantages of joining herds. But Hamilton was referring to protection against wild natural predatory pressures and not *self-induced (artificial) mob-on-mob* type violence that ensues group-think and artificial labels and stereotypes. Masks, labels and mobs are maladaptive for a species that is also prone to self-delusion, addiction and deception.

Masks, Crutches and Daggers Prevent Healing

In my estimate, once we add the costs of our judicial, legal, prison, medical, police and military systems, at least half of our economic growth (GDP, the collective economic value of all goods and services produced domestically) or about $10 trillion relies on masking or fighting *trauma* in our bodies, minds and social relationships instead of healing them. Masks and crutches prevent natural healing because they shield us from natural feedback such as our body's physical and biomarker signals, including Insulin or cortisol overload, hormonal imbalances, weight fluctuations, belly fat, constipation, sexual disorders, fever, sleep disorders, puffy eyes, pale, red or wrinkled skin (particularly neck, back of hands). These disorders all have underlying mechanisms that point to some deep imbalances in the body but in modern times we focus on using crutches to treat the *symptoms* and even medically name the disease after the *symptoms*, such as *erythema* for red skin, *periorbital puffiness* for swelling around the eyes, *prostatitis* for inflammation of prostates, *erectile dysfunction* for impotence, and *jaundice* for yellowness of skin and whites of the eye.

The real disease is the imbalance that causes the symptoms and not the symptoms. For instance, many of our physical symptoms and disorders such as jaundice are signs that our liver may be struggling with detoxifying and cleansing our blood. Puffy eyes, medically termed as *periorbital puffiness*, could indicate several underlying causes such as thinning and weakening of our skin (depletion of collagen and elastin upon oxidative stress and aging), disorders in thyroid hormones or immune system. In the case of visceral (belly) fat, it is also linked to several biological feedback loops inside our body, particularly those including the master hormones insulin and cortisol.

True healing requires addressing the underlying causative imbalances and feedback loops. Trauma is part of nature but so is healing, which always takes time. Remember even hormones, our body's main signaling molecules, act relatively slowly. Our modern approach to resolve trauma, however, is to mask it or use crutches such as cosmetics, surgeries (Botox, laser, liposuction, etc.), creams, synthetic hormones and chemicals. Some of these crutches could indeed cause trauma.

Imagine driving one of those older cars with a chronically overheated engine, like one my father used to drive. Because we could not afford fixing the underlying cause or buying a better car, we dealt with the overheating by going slow on the road and making short trips. What would happen if we covered up the "heated engine" signal light in the dashboard or just carried ice bags to cool down the engine each time we pushed it too hard? And imagine the car manufacturer, instead of improving the engine design, opened garages that specialized in *patching up* overheating radiators and making the car road-ready for another 100 miles. It would be good business for the mechanics, but not good news for the car engine! Yet when it comes to our bodies, we do what we would not do to our cars. We use cheap fuel (food) for our body, ignore its signals, and constantly patch up things by visiting mechanics (doctors). If our body was like a car, it would be painted and patched, pushed to the limit, speeding and overdriving the engine, while carrying a camper full of tools and gadgets to fix it along the way before our next mechanic appointment.

We also know that moderate stresses in nature, a condition called hormesis (described earlier), is helpful in conferring long term immunity on species. So using crutches and tools to shield ourselves from even moderate stresses is maladaptive.

We Become a Fragile, Non Resilient Species

Crutches and controlled environments lead to shrinking of the cortical brain over time. As discussed earlier, studies have shown that wild rats[5] challenged in nature are more resilient and better problem solvers, particularly in times of uncertainty, than lab rats trained with advanced gadgets. Resilience is not about the degree of accomplishment using crutches. It is about surviving and adapting without them

In December 2020, in a boutique hotel in New York City a young woman falsely accused the young son of an acclaimed jazz musician of stealing her iPhone. The young woman later realized she was wrong. When defending herself, she contended that her overreaction was because she really *could not live without that iPhone*! Imagine a world in which a wealthy person with access to all sorts of resources cannot imagine herself without a gadget even for a few minutes. This is a classical definition of a crutch that contributes to our fragility.

A recent shocking report[6] reveals that older millennials (born 1981-1984) are generally and mysteriously unhealthier than not only younger people in the same generation but also Generation Xers of the same age. One plausible and rational explanation is that the oldest Millennials were 13-16 years old in 1997 when mobile phones were rolled out across the U.S. So they are the first generation that grew up *all* their adolescent and adult life *attached* to their cell phones (as crutches). This could mean a more sedentary lifestyle, less independent problem solving skills, and longer and more intense exposure to electromagnetic fields.

A most popular crutch which has become addictive for many of us is internet usage. If you disagree and consider the internet an enhancing tool for humans, you may not be aware of a 2017 study[7], in which the skin conductance of a group of participants were measured before, during, and after an internet session. The galvanic skin response (GSR) and skin conductance response (SCR) are indirect measures of sympathetic autonomic activity, which is associated with both our emotions and attention. In humans, the amplitude of SCRs is related to the level of arousal and anxiety elicited by visual stimuli. The results of the experiment shows that about 27% of users, labeled as problem users, showed increased skin conductance after internet use was stopped, relative to before their internet session. Problem users' GSR scores increased, as the time from internet cessation became longer. They also showed and self-reported increased levels of anxiety, following their internet session, which correlated with their skin conductance scores. Overall, the results suggest that, following termination of an internet session, addiction-type withdrawal-like effects are seen, both psychologically and physiologically. Other studies show that when problem users stop using the internet, they display an increase in heart rate and systolic blood pressure similar to withdrawal symptoms with alcohol and sometimes cannabis.

Chapter Synopsis and References:
Masks, crutches and daggers break natural feedback loops and feed on our vulnerabilities and maladaptive traits, so they beget more masks, crutches and daggers. They self-reinforce our addictions and handicaps, prevent true healing, and make us a fragile species. They make human lives exhausting and metabolically   inefficient.

[1] https://www.newsweek.com/ghislaine-maxwell-demands-arraignment-person-jeffrey-epstein-qanon-live-stream-1581278
[2] Burton, Robyn, and Nick Sheron. "No Level of Alcohol Consumption Improves Health." *The Lancet*, vol. 392, no. 10152, 2018, pp. 987–988.
[3] Kuo Tzu-Hsin and Chiao Chuan-Chin, et al. "Learned Valuation during Forage Decision-Making in Cuttlefish." *Royal Society Open Science*, 16 Dec. 2020.
[4] Yakymenko, Igor, et al. "Oxidative mechanisms of biological activity of low-intensity radiofrequency radiation." *Electromagnetic Biology and Medicine*, 35:2, 2016, pp. 186-202
[5] Welniak–Kaminska, Marlena, et al. "Volumes of Brain Structures in Captive Wild-Type and Laboratory Rats: 7T Magnetic Resonance in VIVO Automatic Atlas-Based Study." *PLOS ONE*, vol. 14, no. 4, 2019..
[6] "The Health of Millennials." *Blue Cross Blue Shield*, 2019, https://www.bcbs.com/the-health-of-america/reports/the-health-of-millennials.

[7] Romano, Michela, et al. "Problematic Internet Users' Skin Conductance and Anxiety Increase after Exposure to the Internet." *Addictive Behaviors*, vol. 75, 2017, pp. 70–74.

Chapter 7: Our Schaub-Lorenz TV Needed a Kick! Good *vs.* Bad Science: The Lobotomy Orgy, the Omnipotence Paradox

"You tell me whar a man gits his corn pone, en I'll tell you what his 'pinions is."
— Mark Twain

Epigraph: Our numbers-conditioned brains create scalable economies and reductionist sciences which thrive on problems (imbalance) and not solutions (balance). This is in contrast to adaptive design principles used in nature based on system wide (wholesome) metabolic efficiency and balance. I will discuss how our modern scalable economic approach has impacted our health and modern sciences.

****

If we were asked to predict the future of a species which (A) Is strongly driven by a concept-driven brain and scalable expansion of its reach, quantity and speed, and (B) Relies on masks, crutches and daggers to adapt, it would be easy to see how we built our current societies and global economies. Our current world is mostly shaped by *Homo economicus*, powerful humans who are strongly driven by numbers and see progress and happiness as scalable quantities. Bill Gates sees prosperity in technology-driven scalable macroeconomic data such as those presented in *Factfulness*. I have reviewed such data, taught Graduate-level university courses in economics, and listened to my late father's anecdotal stories about old time economies. I cannot deny that technology has gifted us with scalable convenience, speed and quantity (income). But *Factfulness* is blind to the fact that happiness is not *exponentially* scalable like technology and the economy. Let me explain.

Neurologically-speaking, as we saw in Chapter 4, what is scale-free is our dopaminergic addictions. The 2010 landmark study by Daniel Kahneman at Princeton demonstrated that happiness indicators correlate positively with income levels up to a threshold (of about $75,000 a year per person in the U.S.) but then plateau and stop rising at higher incomes. Mathematically, they follow a *Gompertz* growth model, an S-shape or logistics type curve that reaches saturation (with time or quantity) as is the case with many other natural growth phenomena (see next page). Furthermore, individual economic gain does not translate to social cohesion, prosperity and peace unless it follows some balancing mechanisms such as those symbiotic distributions we discussed in the chapter on natural evolution.

So if happiness plateaus, why are we so driven by unlimited scalable economic growth? We have built our societies and lives based on a year-over-year growth target of 3-5% which is now common with individual incomes, corporate earnings per share and national GDPs. What if our scalable economic growth has alienated us from natural balances to such an extent that the costs of treating disease, disorder and discord outpaces our scalable gains?

Our desire for scalable growth leads to perennial (current) deficits and the kind of problems I shared in chapter 1. Our societies, roads and hospitals and prisons, our families, friends and social relations, our bodies and minds, and our ecosystem, are all mirrors to our scalable but imbalanced growth. In my upcoming book *Homo economicus*, I will demonstrate with examples how fragile the scalable human economy has become in contrast to the frugal natural (energy and materials exchange) economy. In this chapter, I will focus on our modern scalable *Homo economicus* approach to health and sciences.

Our Personal Balance: Health

Many of us are pickier about the quality of our pet food or gasoline for our cars than our own food. We also know more about how to balance and tune up our cars than our own bodies. Throughout this book, I have equated natural health with a homeostatic metabolic state of balance in our body and mind. Imbalance *is* disease. We saw in Chapter 4 that too much or too little dopamine could make the difference between depression and Parkinson's disease. We saw how imbalances in cortisol, thyroid hormone and insulin were linked to each other and to diabetes, obesity, dementia and depression. Many of these disorders are what evolutionary biologist Daniel Lieberman calls diseases of *evolutionary mismatch*, an imbalance caused by a conflict between our modern lifestyle and our evolutionary physiology and anatomy. In the case of Parkinson's disease, for instance, research shows dopaminergic neurons in the motor control part of the brain, which are uniquely sensitive to degeneration upon disease are large, arborized and energy consuming. These neurons needed to be large in our hunter gatherer ancestors who relied heavily on motor control (physical activity coordination) in pursuit of physical rewards. But over time, our rewards and our pursuits became less physical and not as energy intensive, hence making large neurons somewhat obsolete and tolling. If we don't use it, evolution will lose it. That is why dancing appears to be extremely helpful in improving gait and balance in Parkinson's patients or perhaps in preventing the disease.

An imbalance on the other extreme is also problematic. Too much dopamine is associated with schizophrenia which is usually treated with medicines that are dopamine receptor antagonists. But being diagnosed and medically treated for schizophrenia is shown[1] to increase, rather than decrease, the risk of Parkinson's disease (PD) later in life possibly because of the imbalance and dysfunction caused by dopamine receptor antagonists. As you see here again, the main theme in prolonging and maintaining health naturally (*without crutches*) is staying close to our evolutionary balance and baselines.

As modern humans we are often agnostic about evolutionary imbalances and mismatch that cause diseases. Medical device companies are now developing sophisticated deep brain stimulation (DBS) tools for the treatment of Parkinson's disease, which involves a device similar to a cardiac pacemaker that sends electrical signals through wire electrodes implanted in the brain. Others such as researchers at the Keio University School of Medicine in Tokyo, Japan have been looking at genetic solutions to our cerebral imbalances by creating the world's first genetically engineered monkey to have Parkinson's disease. Other monkeys created as part of the same project mimic Alzheimer's disease and motor neuron disease.

I like to share with you one example of my frustration with reductionist medicine. It was the case of my recurring kidney stones, an extremely painful condition which is likened to the pain of labor when delivering a baby, according to women who have had both kidney stones and babies. The pain is so debilitating that its occurrence in world leaders has been registered as historical events: Roman Emperor Caesar Augustus, Russia's Peter The Great, France's Napoleon, Isaac Newton, Benjamin Franklin, President Johnson, Jack London, and recently Jeff Bezos, all suffered from debilitating kidney stones.

I visited several top urologists in our area but always asked them about the etiology (root cause) of the problem and its prophylaxis (preventive intervention). My questions elicited short and rushed answers in the 5 minutes the doctors could allocate to each patient. One urologist seriously quipped: *"That is why I dislike scientists and engineers as my patients!"* He then suggested I visit him every 3 months for follow ups because it is hard to know what triggers it without diagnostics. Another urologist was sincere enough to acknowledge the limitations of his training in pharmaceutical surgical interventions but not in nutrition or metabolic etiology of the disease. A third urologist, the best known in our area, blankly told me: "You can't keep worrying about what you eat or do. Causation is not easy to establish. I have a cancelation on my surgery roster next week. I can fit you in there; otherwise you may have to wait for another month or so if you can tolerate the pain!" Listening to him, I couldn't help but think about the sales pitch I have heard from car salesmen. Except now I was in agonizing pain so the sales pitch was convincing. I agreed to undergo lithotripsy (shock wave therapy) but I made a pledge to study on my own the root causes of my recurring kidney stones.

Years later, I learned that according to modern holistic immunology, disease (Dis' Ease literally means lack of stability and balance) is rooted in imbalances accumulated in our body and mind over time caused by broken feedback loops which are triggered by our lifestyle. We often receive many signals about these imbalances but ignore them until it is too late and they need elaborate interventions. For example, constipation and/or urinary pH (acidity) may be important early signals (feedback) for some diseases, including kidney stones. If we ignore natural feedback loops and signals (which most of us are not taught about) chronic imbalances lead to acute conditions, which, as with accidents, need immediate medical interventions to mitigate the pain or risk of death.

But kidney stones were only one of my many problems. In my journey of scientific discovery I learned that imbalances lead to imbalances, and diseases lead to diseases. So I pledged to conduct a general etiological study of the malaise in life.

Naturally speaking, hormonal imbalances trigger many of our metabolic diseases. Diabetes and metabolic diseases associated with insulin and cortisol-resistance skyrocketed throughout the second half of the 20th century. It was a double whammy against modern humans. On the one hand, as we discussed, humans binged on cheap (affordable) purified denatured calorie sources such as sugar and high fructose corn syrup. On the other hand, economic motivators and stressors, not being metabolic, overwhelmed the human body's metabolic homeostasis circuits and led to allostatic diseases of mismatch. *Dis'*ease ensued as it always does in nature when there is imbalance and metabolic disorder.

Reductionist Science: Schaub-Lorenz; Microscopic Inspector

What will you get with a concept-driven species that craves numbers, tools and scales? Reductionist science! My fellow curious scientists now dissect everything, even atoms. In many societies, the faith in physical sciences competes with the faith in religions. Yet those worshipping at the *altar of science* also need to be reminded of the serious limitations of reductionist science. For starters, splitting is not always good and some things are not meant to be split because denatured unstable products are costly and dangerous as we saw in Chapter 1. What if the *Omnipotence Paradox* which has been used for God, is applied to modern science? If science is omnipotent, could it create tools that would make the tools, the scientists, and ultimately humans obsolete[i]? I believe the alarming answer is yes.

Let's conduct two thought experiments to see how reductionist science works. First, let's assume we are hiring two inspectors to evaluate a house we plan to buy. Imagine Inspector 1 has no sophisticated gadgets and mainly relies on visual macro signals and markers such as gross (macroscopic) signs of wetness or decay but has a good eye for these flaws in a house. It takes him a day and $1000 to inspect the house for flaws.

Now imagine Inspector 2, relies on advanced microscopes, ultrasound, and even scanning electron microscopes (SEM) to visualize and inspect small defects even at submicron levels on and inside the walls, floors and ceilings. By the time Inspector 1 finishes his inspection, Inspector 2 is still micro-analyzing the floor of one of the rooms because he has seen and documented interesting

---

[i] The original paradox questions if an omnipotent God can create a stone so heavy that even he could not lift!

microbes that *could be* potentially harmful. It takes Inspector 2 a minimum of 30 days to inspect and report, and $30,000 in lab testing to evaluate the house with amazing dissected views.

If you have a limited budget and time, which inspector would you hire or trust? Obviously Inspector 1, the macroscopic one. But what if you had enough money to hire both? What if Inspector 1 approves the overall house condition but the reductionist (detail-oriented) Inspector 2, based on microscopic analysis and costly lab results, observes microscopic fungi that may or may not be harmful so he recommends a disinfecting process that costs an additional $50000 for your peace of mind? So your decision tree includes an approval by Inspector 1 (costs $1000 in total) to buy the house AS IS, or a detailed inspection and treatment by Inspector 2, costing a total of $80,000. Which one would you trust?

Another scenario is when Inspector 2 lobbies the government, banks and realtors associations to mandate *microscopic* evaluation on all home purchases. In that case, you wouldn't even have a choice but spending $80,000 for a report and treatment by Inspector 2. Obviously, that would jack up home prices and benefit not only reductionist Inspectors but also fee-based realtors and mortgage lenders (banks). Such lobbies and mandates would take away individual-level freedom to assess risks and rewards and add a financial burden (external negativity) on home buyers and sellers.

We all have different budgets, timelines, and risk-reward tolerances. If one makes $500,000 a year, he or she would perhaps go with Inspector 2. On the other hand, if one makes $50,000 a year, he or she would view things a bit differently because the elaborate and time-consuming microscopic inspection by itself is a huge cost (risk) and seems absurd for a risk-tolerant person.

The point of this thought experiment is to show you how reductionist science (the approach taken by Inspector 2) can add both value and cost to our risk management strategies. A microscope in the hands of a meticulous inspector often means additional lab tests and follow up which in turn require additional tools that turn into crutches we depend on. Also because our scalable economic growth relies on spending and job creation, reductionist science is often our preferred method to expand our knowledge base despite its costs and long timeline.

Let's go through a second example to demonstrate the limitations of reductionist science. This is actually a real life story from my childhood some 50 years ago. We used to have an old bulky Schaub-Lorenz TV which, like many other old TV models that used the Cathode-ray and electronic vacuum tube technology, occasionally needed a mild kick to get rid of scrambled images. Over time, however, mild kicks did not work anymore and we would call in the local TV repairman. Back then TV repair shops were popular and costly. TV repair, as a specialization, was a well-compensated profession.

The repairman would show up and like reductionist scientists, using his gauges and tools inspect individual bulbs, tubes and connections inside the TV housing cabinet. He would often replace a tube or two, and charge us about one full day of my dad's income to get us a few more weeks of TV watching before the scrambled images, mild kicks and ultimately another repairman visit. This cycle was repeated on a bi-monthly basis. So our family spent about a day of income every other month on repairing the TV through a reductionist, tube-by-tube inspection method.

But one day, tired of my father's gripes about our TV repair costs, I decided to analyze the problem. My approach, as a curious grade-schooler without sophisticated gadgets of the reductionist repairman, was obviously holistic and system-wide like that Macro house inspector in our first thought experiment. I realized there was a lot of heat associated with our TV. Those old vacuum tubes[ii] generated heat and necessitated a lot of ventilation holes on the TV cabinet. In fact, when I peeked through small ventilation holes on the TV's back panel, I would see the tubes glowing in shades of orange when working properly.

So in my grade-schooler mind, the TV's tube circuits performed well like an orchestra as long as their heat-generating instruments had proper ventilation. I asked my father to cut some more holes in the back panel and we regularly cleaned the dust built up in the ventilation holes. To my family's surprise, our TV lasted twice as long in between repairman visits.

So to me, it looked like I achieved, without tools, something the well-paid *reductionist* repairman could not achieve, and with minimum effort and no cost! After a while I figured if the dust and heat are culprits, why not move the TV to a cooler, less dusty part of the room because I had noticed more dust and heat in areas closer to the old heater in our living room. Moving the TV away from the heater and regularly cleaning up the vents added a couple more months to the repair-free periods of our TV. I was proud to help my dad cut our TV repair costs by about 80% with some simple modifications.

What is the point of sharing this thought experiment and real life story? That sometimes sophisticated tools in the hands of a reductionist specialist could distract or cripple him from seeing the big holistic picture, especially if he charges per visit[iii]! Reductionist approach explains *how*, and the holistic approach finds out *why*. Remembering this childhood experience, years later when I chose my PhD research topic I made sure it involved some big picture analysis as well as experimentation and reductionist analytical modeling, to balance the overall reductionist analysis and modeling now adopted by major universities and research centers.

In what follows, I will elaborate on the hallmarks of good and bad reductionist science.

[ii] Tubes acted as voltage-dependent switches, similar to solid-state transistors that replaced them later
[iii] Remember one of the strongest drivers of our brains' self-delusional bias: Self-interest and gain!

Synergism: Reductionist Science and Scalable Economy

There is a strong synergy between reductionist science and scalable economic models. Science allows higher productivity and consumption which demand more tools and scientists to fix the hard working humans and tools. Science keeps us productive and alive, which means higher demand-side pull (consumption) on the economy. Without this scalable consumption, the economic productivity per unit time would have to be capped and there would be no need for additional technologies to increase our economic productivity. In the US, 70% of the economy (GDP) consists of consumer spending, which means demand by individuals. This is why in *Homo economicus* model of adaptation, even the $56000 a year Alzheimer's drug that was recently approved by FDA (discussed in chapter 1) is *economically* adaptive because it may generate a net positive impact on the economy, jobs and GDP. Serving the aged population creates new employment and new consumption.

Aldous Huxley, the savant 20th century author of *Brave New World*, foreshadowed a (fictitious) future in which humans become highly productive yet conformist, perseveratively driven by boundless (dopaminergic) consumptive pleasures, and protected from natural feedback such as pain, anxiety, stress, depression and disease with pills called *soma* only manufactured and administered by alpha humans.

Currently, the most rewarding paths to power and dominance hierarchy seem to be through sponsored science in fields such as information technology and biotechnology. There are many examples of a reductionist approach to solve the world's macro (scalability) problems. A prime example of that is seen in the industrial (scalable) cattle business.

Homogenization of milk was possible because industrial reductionist scientists applied old science, developed by 18th century scientists such as Bernoulli and Venturi, into the new field of fluid mechanics. As discussed in Chapter 2, without the science of homogenization and pasteurization, milk could not be delivered in large quantities, over long distances and at low prices in local supermarkets. This helped humans increase their production and consumption of milk at unprecedented rates. Small farms could not compete with large-scale industrial corporate-owned farms that economically milk cows at the maximum rates possible by (1) Packing a large number of cows in cage-like structures to maximize the number of cows per acre of purchased land, (2) Feeding cows mostly soybeans or corn (often the inexpensive genetically modified corn) instead of grass, (3) Using pesticides and pharmaceuticals such as antibiotics and vaccines to prevent disease in overcrowded dairy pens, (4) Administering to cows hormones such as Bovine somatotropin (bST), also known as bovine growth hormone, to artificially multiply milk production in each cow. But *quantity* and short-term economic efficiency often sacrifice *quality* and long term balances.

This trade-off presented itself in 1986 in the "mad cow disease" scare. The issue was linked by some analysts to commercial feeds used in dairy farms which may contain antibiotics, hormones, pesticides, fertilizers, and even meat and bone leftovers from the slaughtering process, as well as from the carcasses of sick and injured animals such as cattle or sheep. About four million cows were killed during the eradication program in the UK alone.

Another serious problem associated with the unnatural economically driven factory farms happened earlier this century. Starting in 2007, something strange started happening to newborn beef and dairy calves in Europe. They had low blood cell counts, their bone marrow was depleted and their platelets were destroyed. As a result, the animals' blood clotting ability failed and they bled spontaneously from their nostrils, rectum, mouth, and injection sites. There would be bruising on the gums and around the eyes, and the bloodshed in the feces indicated internal bleeding. The condition was usually fatal, at a death rate of 95 percent, with calves dying within weeks of birth. The condition became known as *bleeding calf syndrome*, or as euphemistically labeled by reductionist scientists as *bovine neonatal pancytopenia* (BNP). After evidence emerged that Pfizer's PregSure BVD vaccine was linked to alloantibodies which are transmitted via colostrum to the calves[2], the company eventually had to pull the vaccine from the European market, some six years after the product's introduction in 2004. Later, concerns were raised in New Zealand about whether colostrum or milk from vaccinated dairy cows would affect human health.

You can find countless other problems associated with scalable global farming, published by people and organizations dedicated to alternatives of factory farms and genetically modified organisms (GMO). Examples of such organizations include Organic Consumers Association, Rodale Institute, and The Weston A. Price Foundation.

In spite of environmental and health issues, because of our demand for high quantities and low costs, factory farms expanded throughout the second half of the 20th century, run by mega corporations and enabled by reductionist science. Scientists are now even trying to control the cows' belches and farts by genetically altering cows! As posted on World Economic Forum's Global Risks Report:

"Methane is one of the biggest causes of the problem, after the more commonly discussed carbon dioxide. Given the pure number of cows in the world, farmed for beef and milk, intrepid scientists have spent a lot of time investigating their burps and belches.[iv]"

Good Science is Developed by Good Scientists

According to a story, when Einstein's second wife, Elsa, was quizzed about her knowledge of science and her husband's theory of general relativity, she responded that If the knowledge didn't make her life any easier, she didn't need to know about it! I actually like that practical approach to science, but in today's world, where career scientists are as powerful as politicians and religious leaders,

---

iv The view on the impact of methane is scientifically challenged in recent years because of the transiency of the methane's effect on the atmosphere

we need to have the skills to distinguish good science from bad. As a trained and independent scientist, I like to share with you the kind of science and scientist that I personally follow.

Good science often provides simple unifying theories for complex phenomena. Occam's or Ockham's razor, named after philosopher William of Ockham, is an evaluation principle according to which, other things equal, explanations that posit fewer entities, or fewer kinds of entities, are to be preferred to explanations that posit more. If you remember from the two thought experiments in the opening of this section, the simpler approach to evaluation of the home or broken TV was not the reductionist approach but the holistic one.

An example of how Occam's razor can be helpful is applying it to health-related issues impacting older men. Prostatitis is a term used to describe the condition of men's prostate glands becoming swollen, tender, and inflamed. It can be quite painful, especially during urination or ejaculation. Although painful surgeries and drugs are medically recommended, research shows that a large number of men can simply mitigate the risk of prostatitis through lifestyle factors: adequate (penis) hygiene, reduced lifestyle stress, regular sexual intimacy[3] (ejaculation), safe (monogamous or protected disease-free) sex, a less sedentary lifestyle, regular exercise, adequate hydration (drinking water), a healthy diet including more vegetables (such as tomato) and fruits and less stimulating drinks (tea, coffee, alcohol and soda). So with prostatitis, the Occam's razor for men is to apply lifestyle changes before considering expensive complex medical interventions which may only be necessary for severe acute cases.

Scientists, like other humans, are prone to the self-delusional biases discussed in Chapter 5, particularly those triggered by self-interest and personal gain. *Good* science, the kind which is objective, honest and rational, and not tainted by biases, depends on good scientists who are balanced, interdisciplinary, well-rounded, humble and not bombastic, self-delusional or biased (compromised) by ego, personal gains or complexes. The majority of good scientists are not good marketers of their work, smooth looking, politically powerful or extremely wealthy.

It is easy to note that industrial economies and the evolution of the calculative economic brain in modern humans had a permanent impact on the quality of science and scientists. The brilliant scientists of the early 20th century were often polymaths, a hybrid of a physicist, a biologist, an engineer, a chemist, a philosopher, and often a poet and a musician such as Albert Einstein, Jagadish Bose, Otto Warburg and Albert Szent-Györgyi, who could not be pigeon-holed in any of today's specialty fields defined by grant sponsors. If they were alive today, they probably could not find an academic job. In fact, even back then Einstein (before his fame) had a hard time finding an academic job. They took the time to develop personal relations with other scientists who sometimes rivaled or challenged them. For example, Albert Einstein and Max Planck, besides fiercely debating Quantum mechanics and politics, played music together, with Planck on the piano and Einstein on the violin.

Although job security and prestige may be important to real scientists, they still have deep convictions in the soundness of their theories and purpose in life. This may be the primary reason why Robert Oppenheimer became a staunch anti-war activist when he realized the implications of his monstrous invention. Like Einstein, he was blacklisted by the government but never retracted his antiwar position which cost him lucrative grants and positions. I can't think of many scientists and engineers today who would risk any jobs or grants by being a conscientious objector.

Independent scientists challenge mainstream and authoritative narratives. Einstein had a general contempt for authorities. He once said: *"To punish me for my contempt of authority, Fate has made me an authority myself."*

Good scientists are generally curious and multi-talented. Consider one of the pioneers of cancer research Sidney Farber. Besides his strong science background in biology, he studied philosophy and played the violin. He was one of 14 children in an immigrant family. He was married to a poet and author.

Publishing papers for grants or academic clout was not a key driver for many scientists. Einstein himself was never just a paper-publishing, corporate-sponsored, one-dimensional type of scientist. He started a small discussion group of thinkers and scientists in 1902, named *The Olympia Academy*, in which they spent hours debating broad science and philosophy topics. Their readings included the works of Henri Poincaré, Ernst Mach, and David Hume, which influenced their scientific and philosophical outlook.

Great scientific minds were not *mainly* driven by rewards and grants (dopamine) or competition (testosterone). Consider Thomas Hodgkin, the prominent 19th century physician and pathologist best known for the first account of Hodgkin's disease, a form of lymphoma and blood disease. Hodgkin was born into a wealthy Quaker family who valued education, peace, integrity, service and simplicity. He was not driven by grants or money or fame. Like most Quakers, he was way ahead of his time by advocating for the abolition of slavery and western colonization.

Einstein had been among the first to be invited to the Institute for Advanced Study in Princeton, New Jersey, and offered carte blanche as to salary. To the director's dismay, Einstein asked for an impossible sum: It was far too small! The director had to plead with him to accept a larger salary.

Other cases of good science include collaboration, tolerance and cross-national or cross-disciplinary work that ignore turf or ego wars. For example, combined chemotherapy for cancer treatment was an oncological revolution pioneered by the famous Frei and Freireich team. Emil Freireich came from a poor family. He was outspoken, impulsive and fired from just about every job he ever had! But Emil Freireich was balanced by Emil Frei, almost his complete opposite. Frei was born into a wealthy family, soft spoken and considered methodical. Science historians believe without the *combination* of the two Emils, *combination chemotherapy* would not be possible.

Another collaborative self-balancing team of scientists which revolutionized cancer research consisted of Varmus and Bishop. Harold Varmus came from a wealthy family, studied English literature before studying medicine and was a political appointee to the National Institute of Health. Michael Bishop, on the other hand, grew up in rural Pennsylvania. He finished his elementary education in a two-room school. His father was a Lutheran Minister. Before medical school, Bishop studied music, history, chemistry and molecular biology. He calls himself *a self-confessed book addict*. The Varmus and Bishop team discovered the retroviral oncogenes - cancer-causing gene mutations that are dormant but may be activated by certain retroviruses that can write back into our DNA - and won the 1989 Nobel Prize in Physiology or Medicine.

A more recent example of great interdisciplinary research is work by neurologist physician Martha Herbert who uses a *whole-body* approach to study brain health in autism through metabolic and electrophysiological pathways.

Science today is grossly different from the science of yesterday (most of the 20th century). Although we have a lot more tools, gadgets and experimental data, in most fundamental sciences (like chemistry, physics and math) we have not had as many *game changing* theories as compared to last century. Today's science is more reductionist (specialized), costly, lucrative, corporate-sponsored and driven by data, commercial products and scalable economic productivity than a century ago. Today's scientists may find it hard to imagine that for five days in October 1927, during the fifth Solvay Conference on Physics in Brussels, the planet's most brilliant physicists and chemists gathered in one place to discuss quantum theory principles which are still considered revolutionary today. The leading figures were Albert Einstein, Werner Heisenberg, Irving Langmuir, Paul Dirac, Erwin Schrödinger, Niels Bohr, Marie Curie, Hendrik Lorentz and Max Planck. 17 of the 29 attendees were or became Nobel Prize winners. If you have studied science or engineering, you would recognize most of the names as the founding fathers or mothers of today's science. Most of them never became nearly as wealthy as even today's average top level corporate or government scientists. Many of them had strong philosophical and ideological differences with each other but for better or worse, their love of science united them. Their friendships or differences were primarily not driven by competitive economic motives as is common now among scientists. In fact, Solvay who sponsored the conferences was a wealthy industrialist who valued pure scientific exchange. If he was alive today, he might have been tempted to hire these scientists to develop quantum computers powerful enough to break the cryptographic security that protects cell phones, bank accounts, email addresses and even bitcoin wallets.

Separately, Einstein also engaged in a series of scientific debates, one of the most stunning in the history of science, with physicist Niels Bohr about issues in quantum mechanics such as photons or a spooky (ghost) action in a distance.

In recent years, Kary Mullis, the inventor of polymerase chain reaction (PCR) technique and Nobel prize laureate, exemplifies a great scientific mind: curious, out-of-the-box thinker, multidisciplinary, outspoken, and not driven by grants, mass hysteria, and political orthodoxies. Born into a farming family in rural North Carolina, he went on to study and publish in a wide range of subjects from biology to biochemistry and astrophysics. A surfer and a guitar player, he married four times. He may have been erratic in his personal life and controversial in his views (about the abuse of his PCR invention, HIV and AIDS, and authorities like Anthony Fauci) but his scientific views were objectively driven by curiosity and facts and not biased by money or power.

Unorthodox, Out-of-the-Box Thinkers Make Good Scientists

Good science involves out-of-the-box interdisciplinary thinking and maintaining a balance between reductionist and holistic approaches. Reductionist science is powerful in finding out *how* changes happen but a holistic approach is necessary for finding out *why* changes happen. Many such visionary scientists are not well known today so you need to look outside the mainstream or even Wikipedia (which I find commercially biased) to find them. For example, you can find some of the best objective interdisciplinary biological science in works by Jagadish Bose, Albert Albert Szent-Györgyi, and Otto H. Warburg. If you are interested in the science of how your body works, I recommend reading their works.

Jagadish Bose was a humble genius from India. He was a biologist, physicist and botanist who showed the neurological (cell membrane potential) impact of radio and microwaves even on plant tissues. Today's botanists often do not know much about electromagnetic waves, and physicists who know about microwaves often don't understand plants.

Albert Szent-Györgyi and Otto Warburg were distinguished polymaths and Nobel laureates who (separately) identified some of the underlying mechanisms of cancer long before others, yet their visionary work is somehow hushed or buried.

Szent-Györgyi was the discoverer of vitamin C molecules and much more than a scientist. He was a warrior, peace activist and fighter against fascism. During World War II, he joined the Hungarian resistance movement against the Nazis, and was sent to Istanbul under the guise of a scientific lecture to begin secret negotiations with the Allies. The Germans learned of this plot and Hitler himself issued a warrant for the arrest of Szent-Györgyi who escaped and spent 1944 to 1945 as a fugitive from the Gestapo. I cannot think of many scientists today, the Nobel Prize winning type, who would fight in guerilla warfare against tyranny. In the 1950s, Szent-Györgyi attributed cancer to an electronic (free radical) imbalance and disorder at the molecular level.

Some 10 years before Szent-Györgyi, Otto Warburg had linked cancer to anaerobic glycolysis (metabolism of sugar) by mutagens (agents causing genetic mutations) which are triggered by sugar, smoking, pesticides, chemicals, artificial food additives, electromagnetic radiations (such as radio and microwaves from cell phones or Wi-Fi), and chronic hypoxia (low oxygen)[v]. The unorthodox views of Warburg and Szent-Györgyi placed them at odds with establishment scientists and made them outcasts in some circles. Had the world taken these two brilliant scientists seriously, it would have saved billions of dollars in cancer research funding and millions of lives. Warburg was known to quote an aphorism he attributed to Max Planck: *"Science advances one funeral at a time!"*

Independent scientists are able to explore groundbreaking areas, such as the endosymbiotic theory of Lynn Margulis (Chapter 3), *because* they are *not* part of the established orthodoxy and are free to think outside the narratives. True science grows *outside* narratives, dictates and mandates. Another example of an out-of-the-box thinking in immunology is interdisciplinary works by René Dubos, microbiologist, experimental pathologist, environmentalist, humanist, and winner of the Pulitzer Prize for Non-Fiction for his book *So Human An Animal.* Dubos observed that microbial diseases in general have their own natural history, independent of drugs and vaccines, in which asymptomatic infection and symbiosis are much more common than overt disease:

"It is barely recognized, but nevertheless true, that animals and plants, as well as men, can live peacefully with their most notorious microbial enemies. The world is obsessed by the fact that poliomyelitis can kill and maim several thousand unfortunate victims every year. But more extraordinary is the fact that millions upon millions of young people become infected by polio viruses, yet suffer no harm from the infection. The dramatic episodes of conflict between men and microbes are what strike the mind. What is less readily apprehended is the more common fact that infection can occur without producing disease."

Good scientists are cautious and humble and acknowledge limitations of established orthodox science. For example, despite all reductionist advances in immunology, virology and vaccinology, and the huge budgets dedicated to these fields, honest and accomplished immunologists admit they still do not exactly know what makes exogenous molecules *immunogenic*, i.e., triggering cellular alarm signals from our distressed or injured cells, and eliciting a hostile defense reaction from our body.

In fact, in recent years, unorthodox views of immunology such as the *Danger Theory* have gained support from the likes of Polly Matzinger. Modern immunological theories now view infectious diseases as a result of exogenous viruses and pathogens upsetting the delicate balance (homeostasis) between our body's immune response and these symbiotic viruses. As discussed in Chapter 3, this view is aligned with the "microzymian theory" proposed by 19th century French scientist Antoine Béchamp whose book was banned by the Catholic Church. Contrary to Louis Pasteur's germ theory which blamed exogenous microbes for our health issues, Béchamp blamed diseases on (unhealthy) *imbalances inside* the organism and between the organism and its ecosystem. The

---

[v] Hypoxia could be caused by certain medical conditions, breathing issues or extended use of face masks.

theory has been experimentally demonstrated by the likes of microbiologist Kwang Jeon who has shown that antibiotics that kill bacteria infecting certain amoebae will indeed end up killing the amoebae hosts which have developed a symbiotic balance with the bacteria.

****

I can think of only a few scientists today who would spend hours reading and discussing history or philosophy (like Kary Mullis who recently passed away) or playing a musical instrument. And I do not blame mainstream scientists. As scientists, we are trained, like other humans, to first and foremost think of reward and motivation in terms of the economic value of our time. How would spending hours discussing David Hume increase the adaptive fitness, namely the economic output and income, of the new-age scientist with a high opportunity cost?

In my own science and engineering career, I purposefully varied my jobs to gain *breadth* of experience in R&D, sales, marketing, business development, product development, engineering, training and development, human resources, and technical support. Outside work, I played tennis, table tennis and soccer and studied biology, immunology and history. Yet my colleagues who spent all their time specializing in one product or corporate function, often enjoyed better job security and income than generalists like me.

Specialists are professionals who help us push the boundaries of science. I go to my specialist friends every time I have a question on specific topics like the microscopic science of pulp and paper, superabsorbents used in diapers, or time-released antibiotic medicine. But on interdisciplinary questions, such as those related to lifestyle, nutrition, or the COVID-19 pandemic, I rely on generalists like myself who screen and analyze notes by specialists and look at the larger picture.

Bad Science is Ego-Driven and Tyrannical

Good science is not driven by politics, commerce or egos so it is never tyrannical and never silences debates and dissent by outliers in favor of orthodoxy. New knowledge always builds upon old information by challenging it. Good science relies on collaboration as well as competition. Good science focuses on *why*. Reverse all of this and you get bad science.

Consider ego-driven science like that practiced by William Halsted, the father of radical mastectomy, the breast cancer surgery that excessively removes the entire breast, pectoralis major, lymph nodes near the collarbone and armpit, and even ribs. Halsted was an ego-driven offspring of a very wealthy family, seriously addicted to cocaine and morphine. Despite his flawed personality, he was very influential in scientific circles and continued teaching surgery at Johns Hopkins and promoting radical mastectomy across the U.S. which left thousands of women disfigured with gaping holes in their chests. For nearly a century, medical schools worshipped Halsted.

It is now known that survival from breast cancer is more closely related to how much the cancer has spread *before* surgery than how much is removed during surgery. It's safe to say, Halsted's personal imbalances, combined with his connections and influence, biased the whole medical industry for about a century. Tragically, Bernard Fisher, the iconoclast who showed there was no statistical difference in survival or recurrence between radical mastectomies and less invasive surgeries, was blackmailed and falsely accused of publishing erroneous data by powerful medical authorities. He was cleared in 1997 and victoriously settled a lawsuit he had filed against the accusers.

History is full of stories of bad science in the hands of biased and self-delusional scientific authorities. A horrific example is the use of lobotomy in the treatment of psychological disorders. In the first seven decades of the 20th century medical authorities worldwide performed more than 100,000 lobotomies in which they drilled holes in people's skulls to sever connections to and from the prefrontal cortex. Although the dangerous surgery was presumed to interrupt dopamine signals leading to psychosis, it would also disrupt other critical cortical functions and lead to frequent and serious life-lasting injuries. Despite general recognition of its harms, lobotomy was mandated and imposed on at least 50000 people in the United States, including many children and women, like sisters of President John F. Kennedy and playwright Tennessee Williams. Both women became incapacitated for the remainder of their life. Homosexuals were also targets of lobotomy to make them "morally sane."

The influential American neurologist Walter J. Freeman became a traveling lobotomist. Of Freeman's 3,500 patients, perhaps 490 died and many more were rendered incapacitated. The inhuman procedure was brought to public attention in the 1975 movie *One Flew Over the Cuckoo's Nest* when the main character starring Jack Nicholson was lobotomized as a punishment for his rebellious behavior.

For years, ordinary citizens trusted reductionist science and mainstream treatments like lobotomy. You may be surprised to hear that neurologist António Egas Moniz even won the 1949 Nobel Prize in Medicine for discovering the technique. The procedure was finally banned in Russia, Germany and Japan, and now performed on a very limited scale in other countries. In 1977, the US Congress, during the presidency of Jimmy Carter, created the National Committee for the Protection of Human Subjects of Biomedical and Behavioral Research to investigate allegations that psychosurgery – including lobotomy – was used to control minority and individual rights.

The infatuation of psychologically disturbed scientists and doctors is not limited to lobotomies and radical mastectomy. During the 1920s, the Austrian Eugen Steinach popularized vasectomy for rejuvenation, vigor and sexual potency of the male body. Among the victims were renowned poet William Butler Yeats who, at the age of 69, underwent the Steinach Operation, and claimed it revived his creative powers! In parallel, vasectomy was also practiced in accordance with the science of eugenics which conducted racially-biased castrations in order to improve the human race. Eugenics used a reductionist and quantitative approach to analyze humanity by measuring dimensions of body organs in different races.

The *"ectomy"* (in Latin it means *removal*) orgy continued throughout the 20th century with the popularity of appendectomy (surgery to remove the appendix), tonsillectomy (surgical removal of the tonsils), cholecystectomy (gallbladder removal), hysterectomy (removal of uterus) and many other anatomical carvings. Obviously, some of these operations can be medical necessities but only in life or death situations when underlying causes are hard to fix. The focus of good science should be on *why* and looking into *organismic (holistic) as well as organ-level* root causes. In many cases, removal of an organ which serves an evolutionary purpose as part of a whole organism could cause more harm than good. That was certainly the case with many removals of tonsils and appendices, both now recognized as important parts of our immune system.

Germany under the reign of Nazis also harbored many immoral scientists. An example was Horst Schumann, a scientist, an SS-Sturmbannführer (equivalent to a U.S. Army major), and medical doctor who conducted sterilization and castration experiments at Auschwitz. He was especially interested in the mass sterilization of Jews by using X-ray radiation. Schumann personally chose his test subjects who were always young, healthy, good-looking Jewish men, women and girls.

Bad Science Works for Prestige and Sponsors

By some accounts, the influence of money has irreversibly tainted the scientific process. In *Science Bought and Sold: Essays in the Economics of Science*, editors Philip Mirowski and Esther-Mirjam Sent thoroughly demonstrate the confluence of money and research in modern economies. Other critics have asserted that scientific misinformation has reached such crisis proportions[4] that it is becoming impossible to rationally debate global crises such as pandemics or climate patterns without reliable sources of information:

"The so-called New Economics of Science models scientists as approximately rational actors motivated by nonepistemic considerations such as prestige and salary… Most analyses of misinformation focus on popular and social media, but the scientific enterprise faces a parallel set of problems—from hype and hyperbole to publication bias and citation misdirection, predatory publishing, and filter bubbles.. We aim to dissolve the myth of numbers as impartial, hard, and unbiased."

Even the editor of one of the world's top medical and science journals is now questioning the integrity of some authoritative scientific guidelines[5]:

"It is simply no longer possible to believe much of the clinical research that is published, or to rely on the judgment of trusted physicians or authoritative medical guidelines. I take no pleasure in this conclusion, which I reached slowly and reluctantly over my two decades as editor of The New England Journal of Medicine."

Also recently, Richard Horton, editor of The Lancet, wrote:

"The case against science is straightforward: much of the scientific literature, perhaps half, may simply be untrue. Afflicted by studies with small sample sizes, tiny effects, invalid exploratory analyses, and flagrant conflicts of interest, together with an obsession for pursuing fashionable trends of dubious importance, science has taken a turn towards darkness."

The problem with reductionist science is that it is ideal for customization to the needs of the financial sponsor. That may be why 90% of cancer biology publications, many of them funded by governments or pharmaceutical entities, are *not* reproducible[6]. That does not necessarily mean the research is fraudulent or useless. It just means the researchers used customized settings and procedures to deliver very specific endpoints (such as efficacy of a specific drug delivery pathway) demanded by the sponsors. That is more or less the type of work my classmates and I did for our sponsored Ph.D. research.

Conflict of interest is a hallmark of bad sponsored science. For our new though experiment, imagine manufacturers of energy drinks (containing sugar or corn syrup) and manufacturers of antidiabetic pills offer you as part of your purchase a free state-of-the-art blood glucose meter, one that is easy-to-use and highly recommended on TV and even by your doctor. Would you trust that glucose meter? If you are a critical thinker, you would be suspicious. The blood glucose meter could indirectly benefit sales of the sugary drinks, the pills and visits to the doctor (to refill the pill prescriptions) if it underreports your blood glucose and gives you a (false) sense of satisfaction with the pills, your lifestyle, the drinks and the doctor. If the Glucose meter needs disposable test strips, it could create a perfect feeder system for integrated supermarket/pharmacy chains which sell the pills, the drinks and the test strips. The feeder system expands to insurance companies that pay for the pills and the doctor visits and the test strips if they get reimbursed by the government, for example in the case of Medicare and Medicaid. It's a win-win-win-win-win for the pill manufacturer, the drink manufacturer, supermarket pharmacies, the device and test strip manufacturer, and the doctors. Even you feel like a winner by being able to control your blood glucose while enjoying your energy drinks. Overall, it is a win for the GDP and scalable economy but it is still bad science when a feedback device is subsidized or promoted by parties that have a conflict of interest. It is *not* a conspiracy of evildoers. It's just how sponsored science works efficiently in scalable economic models with feeder systems.

The science of the late 20th century and early 21st century has been shaped by for-profit sponsors. If a business project makes economic sense, the business can always find scientists who will deliver, some even unethically, as was the case in the 1960s when the sugar industry began working closely with nutrition scientists who were paid to single out fat and cholesterol as the dietary causes of coronary heart disease. These scientists downplayed evidence that the consumption of sucrose (sugar) and processed syrups, like high fructose corn syrup, was also a major risk factor[7]. Pediatrician Dr. Robert Lustig has done a great job exposing and chronicling the corruption of science in this case.

Scientists are now in high demand to work on all kinds of absurd projects, such as *Training Spinach Roots as Spies*! Researchers at Massachusetts Institute of Technology (MIT) have engineered the roots of spinach plants with microscopic nanosensors that are capable of detecting nitroaromatics — chemicals that are often found in explosives and man-made industrial chemicals. When the nanosensors detect those compounds, they can send a signal to an infrared camera, which can shoot out an email alert.

In fact, DARPA, the Defense Department's advanced research wing now has an entire new program called Advanced Plant Technologies (APT) dedicated to exploring how plants could be engineered as "spies" to detect threats like chemical and biological weapons and radiation. Here is a quote from the program website:

"Few military requirements are as enduring as the need for timely, accurate information. To meet this demand, the Department of Defense invests heavily in the development of powerful electronic and mechanical sensors, and in the manpower to maintain and operate those sensors. .. Today.. the challenges of monitoring distributed activity are far more complicated. The military's traditional sensors are not always optimal for the task. Fortunately, nature, the master of complexity, offers potential solutions. DARPA's new Advanced Plant Technologies (APT) program looks to seemingly simple plants as the next generation of intelligence gatherers. The program will pursue technologies to engineer robust, plant-based sensors that are self-sustaining in their environment and can be remotely monitored using existing hardware."

As part of this program, for example, the University of Tennessee (UT) Center for Agricultural Synthetic Biology will receive up to $7.5 million. The 4-year effort will combine the expertise of plant biologists, biochemists and engineers. Researchers at UT and MIT are also working to modify plants like potatoes to detect and report potential threats such as nerve agents, radiation and plant pathogens.

Another mind-boggling and spooky research area is biometrics. According to a July 2020 article in *Forbes*:

"Armed with 8,000 employees and an annual budget of between $1 billion and $2 billion of taxpayers' money, Mitre Corp., [is] a government-linked Skunk Works, .. Among the government's wilder Mitre orders: a prototype tool that can hack into smartwatches, fitness trackers and home thermometers for the purposes of homeland security; software to collect human fingerprints from social media websites like Facebook, Instagram and Twitter for the FBI; support in building what the FBI calls the biggest database of human anatomy and criminal history in the world; and a study to determine whether someone's body odor can show they're lying."

Sometimes corporations have larger clout than governments in the science world. Consider the case of smoking. In 1950, Richard Doll and A. Bradford Hill, researchers at the British Medical Research Council published[8] a groundbreaking paper which provided direct statistical evidence that linked increases in smoking in the first half of the 20th century to cases of lung carcinoma and death. This was a few years after Otto Warburg identified smoking as a potential carcinogenic mutagen. It took the world, and even British citizens, some 40 years to hear about Doll and Hill, despite their position with the British government. Even fewer people have heard of Warburg, who also implicated electromagnetic waves. Why? Because scientists and medical doctors, either too busy with their own business, or self-deluded and biased by personal gains, ignored research by the likes of Doll, Hill and Warburg. In fact, many commercial ads used doctors and scientists, as well as actors and celebrities, to promote smoking certain brands of cigarettes. Imagine how many lives would be saved if the majority of scientists, doctors and government officials publicized Doll and Hill's research earlier.

When it comes to clout and funding, even military research is dwarfed by the medical industry[vi]. Medical research has delivered miracles in areas such as regenerative neurogenesis and surgery but it has also been plagued with bad science. David Freedman of *The Atlantic* references research by Stanford University Professor John Ioannidis to document bad medical science in his article titled *Lies, Damned Lies, and Medical Science*.[9]

"Much of what medical researchers conclude in their studies is misleading, exaggerated, or flat-out wrong. So why are doctors.. still drawing upon misinformation in their everyday practice? Dr. John Ioannidis has spent his career challenging his peers by exposing their bad science."

The scientific bias in the medical industry may not be intentional or dark-hearted, just a reflection of a steep hierarchy of influence and power in an industry which, like others, is run by a scalable economic model and focused on steady growth in annual revenues. A total of a few hundred people, the lords of medicine, revolve and recycle in key positions in regulatory agencies and the boards of pharmaceutical corporations, healthcare providers, insurance companies, universities and even news agencies and banks. They are under immense pressure to grow annual revenues and at the same time, minimize liability and risk in an industry which constantly deals with uncharted territories and unpredictable diseases. Growing a $4 trillion industry even by 5% ($200 billion) is not as easy as growing a small family business. Remember from earlier in this chapter, scalable economies rely on deficits and scalable problems to solve but even in our dysfunctional world *new* disorders and diseases may not grow at a rate of $200 billion a year! In wild nature, diseases don't grow at a rate of 5% a year either. If anything, natural selection often improves resilience by rewarding symbiosis.

And in times of emergency, the self-reinforcing biases of human brains (Chapters 4 and 5) kick in. In a shocking editorial published on November 13, 2020, Dr. Kamran Abbasi, the Executive Editor of the British Medical Journal (BMJ)[10], one of the world's most prestigious medical science journals, made these shocking and unprecedented statements about the pandemic policies:

[vi] The annual budget of the U.S. Department of Health and Human Services (HHS), at about $1.5 trillion, is about twice the entire U.S. military budget. Private medical spending is an additional $2.5 trillion a year.

"Science is being suppressed for political and financial gain. Covid-19 has unleashed state corruption on a grand scale, and it is harmful to public health.. Pandemic response relies too heavily on scientists and other government appointees with worrying competing interests, including shareholdings in companies that manufacture covid-19 diagnostic tests, treatments, and vaccines. Government appointees are able to ignore or cherry pick science—another form of misuse—and indulge in anti-competitive practices that favour their own products and those of friends and associates.. The stakes are high for politicians, scientific advisers, and government appointees. Their careers and bank balances may hinge on the decisions that they make. But they have a higher responsibility and duty to the public. Science is a public good ... Importantly, suppressing science, whether by delaying publication, cherry picking favourable research, or gagging scientists, is a danger to public health, causing deaths by exposing people to unsafe or ineffective interventions and preventing them from benefiting from better ones. When entangled with commercial decisions it is also maladministration of taxpayers' money.. When good science is suppressed by the medical-political complex, people die."

Beware of Crutches in Reductionist Science

Remember from chapter 6, crutches are tools that beget more crutches and make you dependent for life. Good science heals and leverages our natural strength. Bad science, however, makes you dependent on a crutch. Scientific crutches include medicines which cause dependence and serious side effects, or genetically-modified organisms that damage the ecosystem (such as genetically-modified mosquitoes being released in some US states to combat mosquito-borne viral diseases such as zika or dengue). A particular category of medicines, which are being scrutinized for their long-term devastating effects, is psychiatric drugs. Groups such as *Mad in America* thoroughly document how the current drug-based paradigm of psychiatric care has failed because many of these drugs are feeders to other psychological drugs and conditions.

The term *neurodiversity* is now used by some scholars to refer to natural variations in the human brain that leads, in a non-pathological sense, to imbalanced conditions in our minds. This non-reductionist approach is becoming popular because it avoids labeling humans with stigmatic psychiatric diseases (such as ASD, ADHD, DCD, ODD, etc.), medicating, branding and isolating them from the rest of the society.

We may all have elements of imbalances that lead to these disorders. By this point in the book, we should have a better understanding that an effective treatment of any disorder should focus on restoring natural balances. Chemical drugs should be only a last resort to restore (forced, partial and temporal) balance. Yet our scalable economy and lifestyles usually encourage the use of crutches, such as chemical drugs, which are often faster and subsidized but leading to dependence and sometimes other dis-ease instead of long-term balance.

Consider Risperdal, an antipsychotic medication given to children *diagnosed* with aggression and anger management issues or autism, ADHD or ODD (oppositional defiant disorder). Risperdal is associated with serious side effects such as obesity, diabetes, heart disease and other disorders caused by metabolic, hormonal and neurological imbalances. So to treat behavioral disorders caused by imbalances which may be common during adolescence, parents and doctors prescribe drugs (crutches) which will make children dependent and may exacerbate their imbalances. And again crutches beget crutches. Good for the economy but not for balance, peace, resiliency, stability and *quality* of life.

Biologists Heather Heying and Bret Weinstein aptly note[11]:

"In combination with restricted access to risk and play (helicopter parenting), and using screens as babysitters, the diversity of legal drugs .. are damaging our children. The considerable rise in mood-altering and behavior-modifying pharmaceuticals being given to children.. is in part a response to children resisting school culture .. Boys are more likely to get diagnosed with ADHD and prescribed speed.. We prefer to drug our children into submission. Girls, on the other hand, .. are more likely to get prescribed anti-anxiety meds and antidepressants."

The authors then posit that neurodiversity is in fact beneficial to many of us because it often trades off one skill with another important skill or interest. It can also break bad educational relationships by misfitting into the generic cookie-cutter canalized education styles in our schools. Remember from earlier in this chapter how (quality) education becomes an orphan in a scalable, quantity-driven, economy and education system.

Unfortunately psychiatric medications are broadly endorsed all over the world to children and adults alike. The market for benzodiazepine drugs is growing fundamentally because of a steep rise in anxiety disorders and broken feedback loops discussed in Chapter 5. The recent COVID-19 lockdowns and short-sighted isolation policies could only exacerbate a state of anxiety and paranoia in vulnerable populations prone to using psychiatric drugs or opioids.

A new study[12] has found that odds of recovery from "serious mental illness" were six times greater if the patient discontinued antipsychotics within two years. The study followed patients with schizophrenia and affective psychosis (bipolar and depression with psychotic features) diagnoses for 20 years.

We already discussed, in Chapters 1 and 4, how benzodiazepines can seriously harm people, as evidenced by the near-death experience of prominent psychologist Jordan Peterson. But benzodiazepines are not an exception. There is a long list of therapeutics that resulted from bad science. Patients trust scientists and the stamp of approval by the US Food and Drug Administration without realizing about a third of FDA-approved therapeutics were involved in some kind of safety event after reaching the market, according to a study published in the Journal of the American Medical Association[13]:

"Postmarket safety events were more frequent among biologics, therapeutics indicated for the treatment of psychiatric disease, those receiving accelerated approval, and those with near–regulatory deadline approval." In simple terms, watch out for medicines and therapeutics that are fast-tracked or just rushed into the market.

You can search, on Google or DuckDuckGo, for keywords like Recalled Medicine or Dangerous Medicine to see the long list including major flops like:

Vioxx (Merck, Causing an estimated 140,000 heart attacks and 88,000 deaths), Bextra (Pfizer, largest health-care fraud settlement and criminal fine in 2005), Fen-Phen (Wyeth-Ayerst, $14 billion awarded to 50,000 victims), Baycol (Bayer, reportedly responsible for more than 100,000 deaths and about as many lawsuits), and Thalidomide (a morning sickness drug for pregnant women, which by 1961 had caused some 20,000 severe birth defects, missing limbs, etc.). According to the National Institutes of Health, this Thalidomide crisis was the "biggest man-made medical disaster ever." In simple terms, it resulted from bad science in the service of scalable economies. By the way, other morning sickness drugs continue to be marketed in the U.S. including one that got Kim Kardashian into trouble with the FDA because she advertised the drug without mentioning the drug's many potential side effects.

Reductionism Needs to Be Balanced with a Holistic Approach

Good science offers a multi-level and multi-scale analysis of the problem, both at the reductionist (dissected, organ or cellular level, microscopes) and holistic (organismic, macro) levels. That is the type of research I conducted for my Masters of Science thesis at Case Western Reserve University. My modeling of certain polymer (macromolecules) blends included linking their (macroscopic) thermal and mechanical properties to microscopic properties analyzed with SEM (scanning electron microscopy) to detect submicron morphological features.

Another example of a balanced approach in science is the modern immunological view of infectious disease, developed in the past 20-30 years. It focuses on *disruptions* to our organismic (i.e. whole human body taken as a system) "*homeostatic balance*" as the source of diseases. For example, the etiology of the mysterious and deadly AIDS epidemic that baffled virologists in the latter part of the 20th century is now explained in terms of an *immunological imbalance* as elegantly described in a 2009 paper titled "*War and Peace between Microbes: HIV-1 Interactions with Co-infecting Viruses.*[14]" This modern view interestingly mirrors some of the holistic wisdom in ancient cultures such as India's Ayurvedic, and traditional Chinese and Iranian medicines

Tibetan medicine, for example, places its primary focus on the interdependence of the body and mind, an ecological view of health, a holistic approach to diagnosing and treating illness. The non-reductionist approach relies on first *relating to* the patient and then *analyzing* their urine (color, odor, viscosity, etc.), pulses (all 12 of them reflecting 12 organs), sclera (white) of the eye, and tongue.

In contrast to ancient traditions, modern science relies on reductionist toolkits but often ignores the importance of multi-level evaluations. A good analogy for the need for multi-level analysis is intelligence and surveillance operations over a geographic area. Satellites can give us a macro view of the terrain, balloons and drones a close up view, and ground intelligence a real-time detailed picture.

The amazing power of reductionist science, like the ground troops in surveillance operations, is its ability to provide us with a detailed (organ or tissue level) view that allows us to connect known dots to see *local and temporal* patterns in uncharted territories so we can venture there with minimal risk in the short-term. But the reductionist view alone, by its very nature, is unable to detect global patterns, and long-distance or long-term risks to us, the ecosystem or the future generations.

One common problem with organ-level drug studies is their lack of focus on organism-level conservation balances, the type discussed for natural systems in chapters 2 and 3. This is aptly highlighted in a paper published by scientists at UK's Medicines and Healthcare Products Regulatory Agency (MHPRA) and the European Medicines Agency (EMA)[15]:

"Despite regulatory guidance and literature publications on the importance of mass balance studies, guidance on how to perform these studies, and the associated metabolite identification studies, it is still relatively common to see deficiencies in these studies in new drug applications. If no mass balance study is submitted or the study is judged as inconclusive, the shortcomings are mainly related to two issues: characterization of circulating drug-related components and elucidation of elimination pathways. In the case of insufficient characterization in human plasma, it is not possible to assess the relevance of the nonclinical toxicity studies in relation to human safety or to fully understand the contribution of metabolites to the exposure response of the entity."

A salient case of organ-level reductionist focus in medicine is immunization. Artificial immunological interventions perform best in goldilocks zones: Too little immune activation leads to infections and cancer, too much to auto-immunity and hyperinflammation. Immune checkpoint inhibitors (ICI), for instance, are used to treat advanced-level cancers by *disinhibiting* the immune system and making it more aggressive in fighting cancer. This novel drug class has significantly increased survival in metastatic melanoma and non-small-cell lung cancer. But it also comes with a risk of inducing autoimmune diseases such as myocarditis[16] (inflammation of heart tissues) resulting from an overly aggressive immune system. This type of dysregulation of homeostatic balances in our immune system is unintentional and hard to avoid if we use a *one-size-fits-all* approach because humans have different physiologies (weight, height, genes, organ size and morphology, etc.) and immune systems (allergies, disease loads, innate and adaptive defense, etc.).

There are also important gender and age differences in our innate and adaptive immune systems. According to a paper published in Frontiers in Immunology,[17] while testosterone is generally found to be immunosuppressive, estrogen is immuno-enhancing, which means females may be less prone than males to viral infections and cancers but more prone to autoimmune diseases and over-reaction to gender-neutral vaccines. Also, because our immune defense is metabolically expensive, as we age, our body invests more in an adaptive immune strategy, which is great in fighting familiar pathogens but may succumb to cancers and new infections.

As we have seen throughout this book, natural evolution and resilience occur *over time* via slow, self-balancing mechanisms but the temporal and local nature of the reductionist approach means it can only focus on short-term solutions. Consider the history of the simian (monkey) virus (SV40) derived from monkey kidney cell cultures, and introduced into human bodies via polio vaccines from 1955 to 1963. Some 36 years later, SV40 was shown[18] to increase rates of ependymomas (central nervous system tumor), osteogenic sarcomas (bone-forming tumors), other bone tumors and mesothelioma (cancer in the lining of the lungs, abdomen or heart) in adults who received these vaccines in early childhood. Unfortunately, Dr. Bernice Eddy, the NIH whistleblower who initially discovered these problems was removed from the task team.

Another issue with the temporal (current time) focus of reductionist science is natural evolution. Artificial immunizations which use inactivated viruses, bacteria or their fragments will be challenged over time by natural mutants and variants of the bacteria and viruses. Remember natural evolution will select for variants that are resistant to selection (vaccine) pressures. This seems to be the case with influenza vaccinations that face a moving target each season in regards to the seasonal variant strains of the virus. A paper published in 2005 finds no strong correlation between increasing flu vaccination coverage after 1980 and declining mortality rates in *any* age group, and concludes that observational studies substantially overestimate vaccination benefit[19]. In fact, the paper observed that influenza-related mortality among the very elderly did not increase markedly during the 1968-1969 pandemic, probably because of persistent immunity acquired through exposure to influenza A (H3) viruses that circulated before 1892.

Pathogenic mutants also escape another group of immune enhancing drugs, antibiotics. These drugs have helped humans in overpopulated and contaminated areas to combat infections. But the overuse of these drugs has turned them into a *crutch* now associated with drug-resistant germs in humans and animals. They can also seriously disrupt our natural balances, gut microbiota and immune system. Research has also revealed[20] that exposure to antibiotics within 14 days of birth is associated with reduced weight and height in boys up to the age of six, and significantly higher body mass index (BMI) in both boys and girls. The impaired childhood growth may be a result of alterations caused by antibiotics in the composition of the natural gut microbiome.

To find more instances of bad or short-sighted science, besides those shared so far, you can research the history of DDT, Zantac, asbestos, Diethylstilbestrol (DES), ESSURE, Endocrine disruptor chemicals and plastics additives, Roundup, Paraquat Pesticide, necrotizing enterocolitis from bovine milk-based preterm infant formulas, or simply look up law firms specializing in handling product liability cases.

Chapter Synopsis and References:

Our economy, society, health and ecosystem reflects our transition from a metabolically efficient species relying on self-balancing natural feedback loops to one driven by scalable economies and self-reinforcing feeder systems. We rely on reductionist science to solve many of our problems. Reductionist science is tool-dependent and helps us connect known dots to see detailed short-term local patterns so we can venture into uncharted territories with minimal risk (short-term). Yet if not balanced by a multi-level holistic approach, reductionist science often misses long-term and long-distance risks to us, our offspring and future generations, and to the natural ecosystems we depend on.

[1] Kuusimäki, Tomi, et al. "Increased Risk of Parkinson's Disease in Patients with Schizophrenia Spectrum Disorders." *Movement Disorders*, vol. 36, no. 6, 2021, pp. 1353–1361.

[2] Deutskens, Fabian, et al. "Vaccine-Induced Antibodies Linked to Bovine Neonatal Pancytopenia (BNP) Recognize Cattle Major Histocompatibility Complex Class I (MHC I)." *Veterinary Research*, vol. 42, no. 1, 2011.

[3] YAVAŞÇAOĞLU, İSMET, et al. "Role of Ejaculation in the Treatment of Chronic Non-Bacterial Prostatitis." *International Journal of Urology*, vol. 6, no. 3, 1999, pp. 130–134.

[4] West, Jevin D., and Carl T. Bergstrom. "Misinformation in and about Science." *Proceedings of the National Academy of Sciences*, vol. 118, no. 15, 2021.

[5] Gyles, Carlton. "Skeptical of medical science reports?." *The Canadian veterinary journal = La revue veterinaire canadienne* vol. 56,10 (2015): 1011-2.

[6] Wen, Haijun, et al. "On the Low Reproducibility of Cancer Studies." *National Science Review*, vol. 5, no. 5, 2018, pp. 619–624.

[7] Kearns, Cristin E., et al. "Sugar Industry and Coronary Heart Disease Research." *JAMA Internal Medicine*, vol. 176, no. 11, 2016, p. 1680.

[8] Doll, R., and A. B. Hill. "Smoking and Carcinoma of the Lung." *BMJ*, vol. 2, no. 4682, 1950, pp. 739–748.

[9] Freedman, David H. "Lies, Damned Lies, and Medical Science." *The Atlantic*, Atlantic Media Company, 2 Sept. 2015, https://www.theatlantic.com/magazine/archive/2010/11/lies-damned-lies-and-medical-science/308269/.

[10] Abbasi, Kamran. "Covid-19: Politicisation, 'Corruption,' and Suppression of Science." *BMJ*, 2020, p. m4425.

[11] Heying, Heather Elizabeth, and Bret Weinstein. *A Hunter-Gatherer's Guide to the 21st Century Evolution and the Challenges of Modern Life*. Portfolio, 2021.

[12] Harrow, Martin, et al. "Twenty-Year Effects of Antipsychotics in Schizophrenia and Affective Psychotic Disorders." *Psychological Medicine*, 2021, pp. 1–11.

[13] Downing, Nicholas S., et al. "Postmarket Safety Events among Novel Therapeutics Approved by the US Food and Drug Administration between 2001 and 2010." *JAMA*, vol. 317, no. 18, 2017, p. 1854.

[14] Lisco, Andrea, et al. "War and Peace between Microbes: HIV-1 Interactions with Coinfecting Viruses." *Cell Host & Microbe*, vol. 6, no. 5, 2009, pp. 403–408.

[15] Coppola, Paola, et al. "The Importance of the Human Mass Balance Study in Regulatory Submissions." *CPT: Pharmacometrics & Systems Pharmacology*, vol. 8, no. 6, 2019, pp. 792–804.

[16] Bruestle, Karina, et al. "Autoimmunity in Acute Myocarditis: How Immunopathogenesis Steers New Directions for Diagnosis and Treatment." *Current Cardiology Reports*, vol. 22, no. 5, 2020.

[17] Taneja, Veena. "Sex Hormones Determine Immune Response." *Frontiers in Immunology*, vol. 9, 2018.

[18] Fisher, S G et al. "Cancer risk associated with simian virus 40 contaminated polio vaccine." *Anticancer research* vol. 19,3B (1999): 2173-80.

[19] "Impact of Influenza Vaccination on Seasonal Mortality in the US Elderly Population." *Archives of Internal Medicine*, vol. 165, no. 3, 2005, p. 265.

[20] Uzan-Yulzari, Atara, et al. "Neonatal Antibiotic Exposure Impairs Child Growth during the First Six Years of Life by Perturbing Intestinal Microbial Colonization." *Nature Communications*, vol. 12, no. 1, 2021.

Chapter 8: Our World is Our Mirror: Paths to Balance, Peace and Health

"The World is Like A Mountain, Our Actions Like a Sound Wave; There is NO ESCAPING from the Echo!"
- 13th Century Persian Mystic Poet, Rumi

Every time I become too critical of the world around me - the media, politics, economy, friends, neighbors and family- I remind myself of this poem by Rumi. Our society, our children, our bodies and health, and our media and politics all just mirror *us*. Unlike other species, human lives are closely intertwined so if I am broken and imbalanced, I will impact many others along my path. Trauma begets trauma. Hurt people hurt people. Imbalance leads to imbalance. One bad apple can spoil the barrel because of the second law of thermodynamics (Chapter 2): Chaos is automatic but balance and harmony need effort.

Thanks to a concept-driven economic brain, we have diverged in the evolutionary path from other life forms on the planet, which naturally evolved as a diverse continuum of species that symbiotically exchange materials, energy and feedback to balance each other. Humans have no particular natural niche. In our convoluted evolutionary path, we have transformed (metamorphosed) from metabolically-efficient nomadic hunter gatherers to surplus-oriented planning farmers, and later to metabolically inefficient cerebral beings driven by scalable economies of speed, quantity and convenience. We have traded off resilience and balance for speed, quality for quantity, and metabolic efficiency for convenience.

We have done away with balance as an adaptive trait yet we spend a lot of our resources to find out about the roots of our trauma. Think of an imbalanced (sick or drunk) person who keeps falling. It is not easy to keep this person up until he regains his self-balancing awareness. Actually, think of an imbalanced or drunk (self-deluded) acrobat walking the tightrope called life.

We have shown throughout this book that it is the principles of biology, psychology, evolution, physics, chemistry and neuroscience that govern our brains and therefore our bodies and behavior. Human-specific trauma, injustice and disease are all rooted in our imbalances. We have lost steady states at all systemic levels: Superorganism (global), colony (social), organism (individual), organ (body parts) and even organelle (cellular) levels. And the problem with imbalance is that we are always on the edge, one stressor away from disease, one paycheck away from bankruptcy, one fight away from divorce, one unrest away from disorder and discord, and one new virus strain away from infections and pandemics.

It took me 15 years to reflect on my life's accumulated imbalances and trauma, including the last two years spent on researching for this book and observing the degree of imbalance and disorder in the human supercolony across the world. So in 2020, when the COVID-19 pandemic struck, I was not surprised at all about the degree of disorder and disease that ensued. Our collective global response to the COVID-19 pandemic just validated how imbalanced and fragile our species has become.

Today, the world is divided, often in a hostile, uncompromising manner, on practically every social, political and economic issue and on how to best solve our problems because we are ignoring how unstable, unnatural and imbalanced we *all* have become. Parents, politicians, teachers and school officials are now fighting over how to open schools and what to teach in them. Even before the pandemic, classical socioeconomic systems such as socialism or capitalism - tried for decades in various countries - had mostly failed to curtail the rampant costs of education, healthcare, pollution containment, natural disasters, wars and crime prevention.

We cannot expect a balanced and peaceful society if its individual members are imbalanced, diseased and traumatized. So balance, peace and health should start at home. And as we learned earlier, the main root of our imbalances are rooted in:

(A) Metabolic inefficiency;
(B) Self-Delusional and Addictive Brain Circuits;
(C) Broken Physical and Social Feedback Loops: Masks, Crutches and Daggers;
(D) Economic Brains Driven by Scale, Quantity and Speed instead of Balance.

Decades of genetic research demonstrates that most human traits are on average 50-60% heritable (genetic). In other words, slightly more than half of our traits and psyche are hardwired in our bodies and brains and chained to our past (nature). But epigenetically, as biologist Bruce Lipton has pointed out, our environment and our proactive conscious control over our thoughts and behavior, can change our gene expressions. So we are like a chain link with 40-50% (epigenetic software) freedom in shaping our traits, which will carry over to the future and become someone else's 50-60% hardware (history and nature). In other words, by reverse-conditioning our bodies and brains we can not only change the 40-50% of our traits that are nurture-dependent but also impact the 50-60% of traits that future humans inherit from us. We can break the chain of trauma and self-delusional brains, and ensure a more balanced, peaceful and healthier future for our species.

The silver lining to our brain's vulnerability to self-delusions and conditioning is our ability to reverse-condition it back onto a path closer to its natural evolutionary roots. In this final chapter of the book I combine science, philosophy and practical personal anecdotes in search for clues about re-conditioning our brains. The content is more or less free flowing and reflects the open-ended nature of the choices ahead of us, as a species which has lost its evolutionary trail in nature.

We Must Become Metabolically Efficient

We learned how our concept-driven brain, taking over our body's driver seat, regularly throws off our metabolic circuits by constantly creating perceived (non-metabolic) threats, rewards and competitive stressors. Yet, as discussed in chapters 4 and 5, recent studies associate diseases such as depression, diabetes, dementia, obesity, as well as infectious diseases and suppression of our immune system, to metabolic disorders triggered by imbalances in hormones like cortisol and insulin.

Research has shown that throughout our life, our hormones *talk* (send feedback in a loop) to each other, synchronize and coordinate their actions via complex feedforward and feedback signals to maintain metabolic homeostasis in our body. For example, growth and thyroid stimulating hormones are both strongly correlated to cortisol levels in our body to meet the changing needs of the organism, such as during the circadian rhythm, sleep, activity, food intake, stress, and inflammation. It is a metabolically intelligent system. For instance, in younger healthy populations, growth and thyroid stimulating hormones correlate positively with stress[1], but in older populations, they have a negative correlation[2]. We also know that with aging or after menopause, levels of several hormones change concomitantly in women because the concerted metabolic impact of multiple hormonal changes is possibly larger than the sum of their individual impact on the aging phenotype.

So to help our hormones and regain our metabolic health, we need to adopt choices in lifestyle, food and thoughts that rein in and train our brains to become metabolically (calorically) efficient again. And I am not talking about extreme or punctuated solutions like becoming a hermit or a forager, or flying to a remote area for a week of fishing and hunting. I am referring to everyday lifestyle changes that *condition* and rewire our brain to think again like the brain of the hunter gatherers. Remember, our brains are neuroplastic and trainable. I am convinced, through my personal experience and research, that our brain and body, like a smart, loyal and trainable pet, can be conditioned with food and lifestyle factors, such as what food to enjoy or detest, or when to eat or stop eating, or when to sleep, go to the bathroom, relax and slow down (we already saw the effect on blood sugar). But conditioning, the same way as training a pet, requires discipline, time and repetition. It will also have trade-off costs. For example, eating healthy food or sleeping or eating on time may require flexible job hours that allow a healthy lifestyle. You may even have to change a job or income source. Conditioning for metabolically efficient and healthy life is just a matter of prioritizing it over other factors in life, even our income and relationships.

Eating the Natural Way

Today, an average kilocalorie of food that we eat, consumes 10 kilocalories of energy (fossil fuel, etc.) to produce, process and ship[3] unless it is *seasonal* vegetables grown in our backyard or local farms requiring minimal processing, storage and transportation. In addition, we waste a lot of energy in the packaging of food. Most people do not realize that the amount of energy used to make the aluminum in two cans of soda is the same amount of calories (about 2000 kCal) needed by an average human to survive for a day. Everyone focuses on the calories inside the can but what goes into making a can or plastic bottle is often a lot more. An average American uses a can of beverage a day so in addition to calories we consume as food, we waste about 50% more just in buying aluminum cans.

To exacerbate our negative metabolic impact on the planet, we often eat when we are not really hungry. Constant overeating induces brain insulin resistance even before peripheral insulin signaling is impaired, implicating *brain* insulin resistance as a key culprit of metabolic disease and diabetes[4]. Insulin crosses the blood–brain barrier and binds to insulin receptors widely expressed throughout the brain. Insulin signaling in the brain regulates appetite, adipose tissue lipolysis (breaking down of fat storage or what we called the catabolic mode), hepatic triglyceride secretion and branched-chain amino acid metabolism, protecting the organism from ectopic lipid accumulation (too much fat build up in liver and muscles that leads to insulin resistance) and toxicity. In other words, unless we eat only when we are *catabolically* hungry, our brain *will* ignore insulin and our body *will* accumulate fat and resist insulin signals.

Remember the four main metabolic modes of our body: Break Down, Bury, Build and Burnish. Our *break down* cycles are mainly triggered by cortisol, the catabolic hormone which breaks down our fat and glucose storage in times of stress or in anticipation of our body's energy needs, such as before we wake up in the morning. Now if we have some extra fat we want to get rid of, and our morning and daytime jobs do not use up much of the sugar released by cortisol, we are better off allowing cortisol to use up some of our stored fat and glycogens before we eat more calories (and activate insulin). This is the concept behind time-restricted eating, which limits the eating window to 8-10 hours a day, for example between 11 AM and 6 PM, and may involve skipping breakfast (by combining it into an early lunch or brunch) or late dinners. We will feel real catabolic hunger hours *after* our *burn* cycle kicks in and we start feeling a bit weak and craving for food in our entire body, particularly the digestive tract and our mouth. Now if someone has a physically demanding (calorie-consuming) job or a lean body, they may not benefit much from time-restricted eating. As with other one-size-fits-all diets and routines, time-restricted eating windows are not for everyone. What is important to realize is the wisdom behind the routine which is turning our body to a catabolic mode if we have excess fat or an inactive life and to train our brain to recognize when to eat by following our body's physical, catabolic hunger.

Certain *ketogenic* diets, that are rich in fats and low in carbohydrates can also trigger a catabolic (fat burning) mode in our body and ketosis (formation of ketones) by suppressing insulin and raising glucagon (the catabolic hormone that balances insulin). But here again, balance should be our goal. Too much catabolism, which already happens with diabetes (insufficient insulin) or with carbohydrate starvation may lead to ketoacidosis which is harmful. I personally use a low carbohydrate diet balanced with fibrous vegetables, protein and fats but avoid extremely high fat or high protein diets. Carbohydrate starvation and ketogenic diets are shown to create unfavorable metabolic environments for cancer cells and used in combination with standard chemo- and radiotherapy treatments.

Eating only when *truly* hungry, exercising, low carbohydrate low protein ketogenic diets, and other temporary physical stressors such as sauna or cold baths, also trigger a catabolic (cortisol-induced) state of *autophagy* which literally means "self-eating" and allows the body to cannibalize waste products, toxic proteins and possibly cancerous cells when occasionally starved on exogenous amino acids and carbohydrates. Again, achieving balance should be our goal[i] so excessive fasting and stressing the body could break down our healthy muscles and tissues which are particularly valuable in aging populations.

Eating less and consuming fewer calories may actually trigger weight loss better than physical activity and exercise can. Research shows differences in obesity prevalence between hunter gatherer and Western populations correlate primarily with differences in their energy intake rather than total energy expenditure or level of physical activity[5].

Once you train your mind and body to eat only when really (catabolically) hungry, you can even correlate your weight gain and loss to your physiological hunger and appetite. I have noticed my fasting body weight (after waking up and urination but before drinking or eating) closely tracks my catabolic physiological hunger the night before. If I feel 95-99% full when I stop eating for dinner (around 6 PM) I keep my steady weight the next morning regardless of how much or what I eat[ii] in my 7-hour feeding window. For exercise, besides the push lawn mower (in warmer months), and walking and aerobic cardio (elliptical) machine, I burn about 1000 kCal (kilo calories) a week, high enough

---

[i] To prevent insulin resistance and diabetes, people who do not curb their eating and carbohydrate-rich diets or balance it with physically active lifestyles, may have to resort to medicine that would regulate glucose metabolism. Another option is to balance sweet foods with bitter food like bitter melon which contain Quinine-like compounds that regulate glucose response.
[ii] Except for extra salty or sugary food which can lead to water retention and weight gain

to burn the desserts I enjoy but low enough not to induce too much catabolic stress and hunger.

Another component of metabolic efficiency, besides reducing psychological stressors in life (described later), is to use food that adapts us to stresses by regulating the stress-induced inflammations or boosting our immune system. Herbs and adaptogens such as turmeric, ginger, ginseng, garlic, onions, astragalus, ashwagandha, reishi, holy basil, fennel, arctic root (Rhodiola rosea) are potent immunomodulators. But herbs should be used cautiously and under supervision of an expert because they are naturally potent. For example, ashwagandha may increase the sedative effects of benzodiazepines such as Valium, Ativan, Xanax, and other depressants.

It has been shown that the beneficial stress-protective effect of adaptogens such as rhodiola rosea (arctic root)[6] is related to creation of a state of *heterostasis* by inducing a slightly elevated level of nonspecific resistance to stressors and decreasing sensitivity to them. As a result, adaptogens have a mild prolonged stimulatory effect (vigilance, energy) on the body and reduce both catabolic stress overload (blood glucose overload) and the exhaustion and fatigue that ensue afterwards. Many adaptogenic herbs have chemical structures similar to adrenaline, which is a mediator of our sympathetic nervous system. Certain breathing techniques, like one promoted by naturalist Wim Hof, can also induce a sympathetic release of adrenaline in the body and act adaptogenic in stressful conditions. Again, like all other hormones, excessive or prolonged release of adrenaline could cause disruptions to our health.

Nutritionally, most food in its natural form, not denatured by processing (as explained in chapter 2), if consumed in moderation and when our body really craves for it, should not disrupt our metabolic homeostasis. Protein, for example, in natural meat form, is diluted and balanced by fibers in connecting collagen and gristle tissues, and by fat which is metabolized in our body by acetyl carnitine in meat. Oranges and apples are packaged by nature with a lot of fiber, such as pectin, which slow down the release of fructose into our blood. Pure human-made concentrated forms of denatured food such as amino acid protein powders and fruit juices will not be as naturally and efficiently metabolized by our body. A widely used denatured concentrated drink is alcohol. Our body really doesn't know what to do with concentrated alcohol, so that is why alcohol, as a fuel, initially causes vasodilation (relaxation of blood vessels), elevated metabolism, energy and sense of elation, followed at higher levels and time exposures by vasoconstriction (shrinking of blood vessels), headaches, low energy and hangover. The short term inhibitory effects of alcohol such as elation and reduction of anxiety, lead upon withdrawal (reduction of consumption) to an increase in sympathetic activity (nervousness) and therefore, dependence and addiction[7]. Alcohol is not available in nature at high concentrations except as a toxin to stem and progenitor cells, i.e., life.

Generally speaking, a metabolically efficient diet is one closer to its natural form with minimal purification or processing. There are only a few natural processes which make food more metabolically efficient, such as slow fermentation.

A group of naturally processed foods that seem to *microbiologically* make us metabolically efficient, and therefore less prone to metabolic and infectious diseases, are fermented foods such as yogurt, sauerkraut, kimchi and kefir. As referenced in my website and social media blogs, we do quite a bit of fermentation with lactic-acid bacteria in our household using locally sourced, raw organic ingredients. Store brand fermented products limit the degree of fermentation because the ongoing anaerobic process could release too much carbon dioxide, which acidifies and destroys harmful pathogens but can also cause dangerous pressure build up in pre-packaged store brands. This is perhaps why store-brand Kombucha bottles are often made out of thick glass. Our home-made fermented foods are often sour, indicating the formation of a good amount of lactic or acetic acid which feed the beneficial probiotic microorganisms but prevent against most other microbes. Fermented foods such as kefir are now labeled as *psychobiotic* because they are shown in animal studies to signal through the microbiota-gut-immune-brain axis[8] and positively modulate host behavior (oxytocin, social bonding, serotonergic activity) and immune system.

This is not a book on nutrition. There are a lot of papers, blogs, podcasts and university programs on nutrition but unfortunately nutrition is not meaningfully incorporated in most (pharmaceutical-sponsored) medical schools. It is greatly advantageous if your doctor is well-educated on nutrition science or a graduate of an integrative, functional, naturopathic or alternative medicine program such as those offered at University of Bridgeport, University of Arizona, Virginia University of Integrative Medicine, Bastyr University, Canadian College of Naturopathic Medicine, Southwest College of Naturopathic Medicine & Health Sciences, National University of Health Sciences, National University of Natural Medicine. Currently, about half of the states in America and most provinces in Canada license naturopathic doctors. This is great news because studied separately (not integrated), nutritional and medical sciences can be reductionist and commercially driven by pharmaceutical or food industries.

Like everything else, our diet (and if needed, *together* with medicine) should make us more metabolically efficient and homeostatically balanced. So instead of one-size-fits-all diets, people are better served by monitoring their own biomarkers to decide which diet suits them best. I regularly screen different sources on nutrition and include my insight and analysis on my blogs.

Living a More Natural Way

To become a more metabolic organism, like our distant ancestors, we need to create physical challenges for ourselves throughout the week. They don't have to be extreme or high intensity. Energy efficiency can start from our households. Our home thermostats are set close to 80 'F for two warm months, and 64 ʼF for five colder months in our region. For about five months a year, we use neither heating or cooling in our home, by just leveraging tree shades and managing flow of natural outside air, occasional window fans, window shades, blankets and proper clothing layers. In the age of virtue signals, my family and I often boast about the virtue of conservation, and we try to practice it. According to our local power company's statements, our household's average energy usage is about half of an average American household in our climate zone even though we live in an older home with mediocre thermal insulation materials and design. Living in smaller, more energy efficient homes and withstanding some cool and warm temperatures can not only save us money on energy bills, it also makes us individually more resilient and metabolically efficient, by wasting less body energy, eating less and being exposed to thermal hormesis, which is believed to be naturally beneficial for boosting the immune system.

We must spend more time outdoors and in nature throughout the year, preferably every day. Natural adjustment to local seasons could also mean less food and physical activity, longer sleep hours, and more indoor family time in winter months.

The lifestyle factor that impacts our diurnal cortisol rhythm and metabolic efficiency the most is our sleep and stress (wake/work/think) schedule. Flatter circadian cortisol patterns, as already shown, are associated with metabolic dysfunction, depression and disease. Studies have demonstrated that depressed patients with disturbed cortisol rhythms might benefit from restitution of those rhythms but patients with more generally elevated cortisol levels might benefit from cortisol blockade[9].

Even the benefits of fasting and time-restricted eating, referenced in the last section, seem to be related to regulating our circadian cortisol cycles. Recent research has shown[10] that fruit flies exposed to time-restricted feeding (fasting) experience lifespan extensions ranging between 10% and 50%, but only if they have a functioning circadian (sleep/wake) clock in internal organs. Flies exposed to constant light or those whose neural circadian clocks have been abolished genetically, did not benefit from fasting.

Research shows that life forms ranging from bacteria to humans are programmed by circadian clocks that are a coupled (coordinated) network of cell and tissue clocks in each organ, receiving their input signals predominantly from cues such as light and food and converting them into timed functional outputs, which, in turn, act as inputs to other organs and effectively connect the body's circadian clock network[11]: "*Circadian misalignment, wherein eating and sleeping patterns oppose their natural inclination from the light-dark cycle, disrupts homeostasis and leads to internal imbalance—a feature of diseases ranging from metabolic syndrome to cancer. By contrast, proper alignment and internal synchrony have been demonstrated to combat tissue dysfunction and promote well-being.*"

Temporary or long-term moves across a wide span of latitudes or longitudes will also disrupt the time or duration of daylight, respectively, and therefore our circadian alignment and metabolism. Disorganized lifestyles associated with unpredictable circadian cues and stressors such as food and artificial light can also disrupt our metabolism.

One way to be metabolically closer to our ancestors is to mimic their lifestyles. For example, waking up naturally (without alarm clocks) puts our body back in charge of our metabolism (as opposed to our *Homo economicus* brain worrying about all types of deadlines and goals). We also already discussed the benefits of occasionally walking or jogging barefoot. Mimicking hunter gatherers simply means walking, climbing and occasional running, preferably in nature, but if not possible, on exercise machines, and breathing through our nose, as much as possible. Mimicking our farmer ancestors could be as simple as mowing the lawn with a push mower (if you have a lawn), like I do. Many of my neighbors spend $2000-$3000 on a lawn tractor and $200-$300 a month between gym membership fees and hiring landscapers. They could save all of that expense by mowing their own lawn themselves using a push mower.

Driving cars is a major root of our departure from metabolic efficiency. A 170-pound person often carrying around a 4000-6000 pound externally-fueled vehicle (air-conditioned shell) is very unnatural. Evolutionarily, we were *born to run*, as aptly noted by Christopher McDougall in a book with the same title: *Born to Run,* which we already discussed in relation to the harm caused by thick-soled modern shoes and benefits associated with barefoot walking or jogging. An internal combustion engine is at best 25% efficient in converting fuel calories to useful mechanical energy (that's why cars have radiators to dissipate the remaining 75% converted to thermal energy). Human and animal muscles also often operate at a maximum efficiency of around 25%. But because of the elastic properties of our tendons and muscle, when we hop, run or bike, our muscles act as springs, transferring the stored potential energy into kinetic energy. As a result, the metabolic efficiency of a running human is nearly 50%[12], more than that of hydrocarbon-fueled combustion engines. Our most efficient muscle in the body seems to be the first dorsal interosseous (FDI) muscle of the hand with an efficiency of 68%. The muscle is responsible for lateral movement of fingers and is perhaps an evolutionary indicator that we were also born to hunt (hold spears, rocks, bows and arrows). So one can say we are evolutionary palimpsests (altered artifacts still bearing visible traces of its earlier forms) of our ancestor and have inherited their efficient human-specific grip for shooting, grabbing bows and arrows, tools and utensils, and in recent years, holding pencils and text messaging! Without our super-efficient dorsal interossei, we probably could not easily adapt to the use of mobile devices and text messaging. Sometimes I wonder if our hunter ancestors could imagine the new use of these strong interosseous muscles.

How to Reverse-Condition Our Self-Delusional Addictive Brains

As discussed earlier, our concept-driven metabolically-inefficient brain makes us prone to self -delusions and all sorts of biases and dysfunctional social behavior. The hormonal soups released in response to scalable (boundless) economic rewards, threats and competitions, condition us to reinforce our addictive habits. Today, at least in Western societies, surviving for many people means a mental state of constant stress, pursuit (of income) and competition with coworkers, neighbors and even with spouses, friends, siblings, and strangers on social media.

We may not be far from Huxley's world of boundless dopamine-releasing pleasures. Mega events such as Lollapalooza and Electric Daisy Carnival push the boundaries of human imagination and attract millions of fans by offering cutting-edge sound, light and color, and often a rave atmosphere of party, ecstasy and psychedelic drugs that would indulge modern humans' dopamine-seeking minds in ways that our ancestors could never dream of. Our pampered dopaminergic brain circuits have certainly come a long way. For our early hominin ancestors, the sight, taste and calories of a bunch of figs would be totally motivational (dopaminergic) to live and fight for. For the indigenous Lenni Lenape tribe who in 1626 sold Manhattan island to Dutch settlers for 60 guilders (about $600 in today's gold value) the Wampum (shell beads) offered as collateral of the trade were more motivational (dopaminergic) than the money. The Lenni Lenape did not believe in the ownership of land because they surmised the land is a living entity so cannot be an object of ownership. Some of the current inhabitants of that land (Manhattan) have evolved brains that will not release dopamine for anything less than world dominance in stocks or political power.

We may no longer be able to think and live like the Lenni Lenape but based on what we learned in Chapters 4 and 5 about brain circuits and hormonal soups, there are ways we can reverse-condition our brains out of habitual ruts and addictive lifestyles:

Slowing Down and Slimming/Scaling Down

When we step away from large or fast rewards and learn how to be happy again with small *amounts* and *rates* of rewards, as we were in our childhood, we train our dopaminergic circuits to *trickle* dopamine. This is not about dropping our long term or ambitious rewarding goals in life. It is about training our mind to become flexible and relaxed enough to feel the pleasure of small inexpensive rewards and experiences. This happens by intentionally managing our reward expectation baselines and set points. For example, do not wait for the perfect week-long vacation to travel to a remote fancy resort. Take advantage of smaller windows, half or a full-day at a time, of great weather to enjoy local pleasures such as natural preserves or state parks and forests. These are less commercialized and marketed so lesser known, more affordable (often free) and less busy. You just need to spend some time finding them. I list some of the best gems we have found, including some with amazing views and waterfalls, in my blogs. Many of these natural getaways are within a 1-2 hour drive for most people so there is no need to spend a lot of time, fuel and money in transit, lodging, restaurant food, etc. You do not need to buy pricey fancy gadgets and outfits or plan and save months in advance either.

We can dilute dopamine release by conditioning our brains to seek smaller, less fancy, less costly pleasures. Avoid being wowed! Discouraging *surprises* and lump sum big rewards minimizes dopamine binges and addictions. We don't want to end up like many lottery winners, professional athletes, entrepreneurs or young Hollywood actors whose simple life and pleasure baselines are messed up when they hit big jackpots. We don't want to use strong stimulants and dopamine releasers like drugs, sugar, and alcohol, which are often pure extractions, hence denatured. The goal is to *slow down* and *scale down* both the pleasures and stresses of life. It has been shown that even in simple life forms without dopaminergic systems, firing of neurons in pursuit of reward (such as mating and reproduction) acts as an intrinsic sub-circadian biological clock. For example, a single neuron in male fruit flies can keep track of time in a silent manner to time their climax during mating at exactly 7 minutes[13]. If like fruit flies, our body calibrates its clock by timing of essential habitual rewarding experiences (food, money, entertainment or sex)[14] the more we space dopaminergic activities, the more we can slow down our internal clock calibrations.

Slowing down our internal clock is advantageous because aging and metabolism are regulated by our perception of time. As explained in Chapter 5, our *perception* of time is shown to control our glucose metabolism and vulnerability to diabetes. To slow down our aging and our blood sugar spikes after we eat, being exposed to fake clocks that work slowly is not practical. But we can slow down our perception of time by slowing down our dopamine and cortisol signaling, in other words, our pace of life.

Slowing down in life has neurological as well as metabolic benefits. Remember how our amygdala and insula, when faced with urgency and short time-to-react, do not wait for input by the prefrontal cortex and use biases, prejudices or pre-programmed anxieties in making decisions (for us)? Several experiments have revealed that when people are shown pictures of unfamiliar faces or environments, they show more bias and prejudice if pressed for a quick reaction, but more judicious and rational when allowed time for feedback. Neurological and hormonal feedback loops need time. Without slowing down in life, our decisions will be more amygdala-driven than cortical. My own family's experience showed me that once I left my corporate jobs and the fast lane, dealing with fewer deadlines allowed me to have more *time* and *energy* to listen to, and communicate with, my family and to solve problems more effectively. What my family lost as the opportunity cost of slowing down my career ambitions, we gained as a more peaceful, less anxious and better problem solving household, avoiding a whole lot of cortisol overload!

Among ways to slow down our perception of time are deep breathing techniques and socializing with people who are not subject to time-sensitive deadlines and scalable economic models. I know goal-driven people who remind me of my own youth, with fast internal clocks. They move from project to project. They are always anxiously seeking the next reward or challenge (dopamine). Because their subjective perception of time spent on a task or pursuit of reward is more than the actual time spent, they demand larger rewards or reach the temporal discounting state (impatience) faster than others or start new challenges. The worst thing that one can do for these folks is to socialize with them and place them among other competitive impatient achievers. Now the *subjective* value of rewards they seek is diminished not only by temporal discounting (via their fast internal clocks) but also by their comparative social benchmarks. These people often end up with superb economic productivity but also diseases associated with metabolic dysfunction. They are socially and metaphorically equivalent to the *Super chickens* discussed in chapter 3.

I believe our *perception* of time is filtered down to cellular levels and even felt by our genes (DNA). This may explain the so-called *Hispanic Paradox* which refers to Hispanics in the United States having a longer life expectancy than Caucasians despite having a higher burden of cardio-metabolic risk factors. New studies that track genome-wide DNA methylation levels[15] as epigenetic biomarkers of aging at the cellular level, show Hispanic populations have slower epigenetic clocks than Caucasians despite higher levels of inflammatory risk profiles. In other words, Hispanic populations have managed to slow down their internal clocks at the molecular level so their genes and bodies are aging slower than their fast-paced, anxious neighbors who may even be leaner and do more cardio exercise.

Rewiring Our Brain's Risk-Reward Calculation Algorithms for Quality

We have turned into computing monsters. Our brain uses only 10 watts of power, the calories in 3 bananas a day, to do up to 20 quadrillion calculations (synaptic firings), which is one million times more than a modern computer. But we still easily become self-delusional and are bad decision makers because *qualitative* aspects of our lives cannot be *computed* using our economic efficiency benchmarks which drive most of our decisions. In Adam Smith's capitalism, a person's unit of time (labor) is valued and quantified by the amount of labor productivity of that person in that unit of time. Unfortunately, many people have now adopted this market valuation of hourly labor (salary) as the baseline of how they "value" their own time and life's worth. If the reward (economic) of our pursuit is larger than expected, as benchmarked against our time's opportunity cost we pursue it. For example, a nurse whose time is valued by the medical market at $30 an hour would lose about $240 if she took a day off work. A medical doctor making about three times as much as the nurse, would value her time in life at a rate of $720 per 8 hours. That is why with time-consuming chores such as mowing the lawn, taking care of an elderly parent, or even educating one's own children, it is more likely that the doctor, and not the nurse, would outsource (hire help). The opportunity cost (i.e., forfeited income) for the nurse, of not getting paid for hospital work while nursing a child or an elderly parent at home, is less than her cost of paying someone else to nurse her child or parent so she would provide the care herself. For the doctor, however, the economic calculus is very different. She comes out way ahead in terms of the opportunity cost if she can work on that day (make $720) or not use a paid vacation and instead pay someone about $240 to nurse the elderly parent or $100 to landscape and manicure her lawn.

The problem with valuation of rewards in terms of their *quantity* or *size* -- as opposed to nutritional and reproductive qualities valued in nature, or simple life experiences valued by early humans and children -- is that it can lead to addictive or risky behavior. I already shared the story of Clifton Maloney, a wealthy investment banker, former VP of Goldman Sachs, and husband of U.S. Congresswoman Carolyn Maloney, who seemed to have it all, money and political power. Yet he was driven (dopaminergically) by the height and number (quantity) of tall mountains he climbed (conquered). In 2009, he died while pushing his luck one last time to climb the world's sixth-tallest peak, Cho Oyu in Tibet.

A scalable (quantitative) type of risk-reward calculus, if not caloric (for metabolic homeostatic budgeting), is evolutionarily foreign to humans. We can recondition our brains by thinking in terms of *quality* of rewards instead of their quantity, speed or opportunity cost. Examples of quality rewards which are time-consuming and therefore high (opportunity) cost include raising good children at home, making home-made food, fostering peaceful, equitable and balanced relations with spouses, children, neighbors and coworkers, and most importantly balancing our own body's metabolism (health) and educating ourselves with objective facts. For many high-income people, it is more economical and less exhausting to *pay* for food, health and education than spending time on them. The rewards we reap from personal time and energy spent on values we cherish in life (health, peace, knowledge) are long-term and qualitative so our economic (*Homo economicus*) brains cannot properly assess them. Maybe that's why humans tax material rewards but not experiential joys of life like a nice walk in the park or raising good children. There are also no easy ways to assign an economic value to what we learn or feel through our life experiences. Many people realize this only in hindsight when the experiential opportunity is gone.

In the age of *Homo economicus*, in my personal relations, I equate *caring* with spending *time* (not money). Instead of *quantity* of relations, my family now focuses on the *quality* of our relations with people who *care* enough to spend *time* with us. Also, in our job or project choices, we factor in quality for instance by choosing projects which are not unhealthy for our body, mind, and conscience. That is why in recent years I have chosen to work on space and NASA-related grants instead of weapons programs, and stayed away from lucrative jobs in alcohol, chemicals and tobacco industries. These decisions have cost us in *quantity* but not *quality* of life.

The problem with economically-efficient decision making is that money-time trade-off for happiness is not linear. If it was, rich people would never pay for love, get depressed, addicted or suicidal in an economic utilitarian system. The visionary coder, mathematician and entrepreneur turned anti-government activist, John McAfee, said later in his life that love, compassion and caring for others do not need economic or political power. He believed power instead fuels corruptive traits like greed, hostility and jealousy.

McAfee may have had a point. In my own personal observation of numerous end-of-life epiphanies by friends and family members during peace and war times, I was surprised to learn that almost all dying people, rich and poor alike, wished in their lifetime, they had lived a less stressed, more peaceful and loving life, had enjoyed simpler things as they did in their childhood, and had become wiser by learning more about *why* things happen. I don't recall anyone who wished they had managed their opportunity cost of their life on the planet better. Regardless of how economically efficient our brains have become, we are still carrying those bodies and souls (affections, memories, empathies) that need to be metabolically balanced and healthy. In *Dictionnaire philosophique*, Voltaire uses a quote from Confucius to define a virtuous and just life: "*Live as you would wish to have lived, when you come to die.*"

Rewiring our brains for valuing *quality* of our time means assigning high values to *simple inexpensive non-competitive experiences in life that take time, personal care and patience* such as arts, cooking, nature activities, raising children, animals, plants or trees, caring for others (elderly, parents, strangers), nourishing our brains and developing friendships. Expensive or exciting experiences such as shopping or adventures do not count because they fire up our brain's dopamine binging circuits and quantitative calculators again. The real rewards of time spent on quality would be measured in improving our state of *balance* as judged by certain biomarkers shared later in this chapter. In its decision making and "opportunity cost per unit time" calculus, a quality-conditioned brain considers all costs (not only quantitative economic ones) including metabolic (allostatic) loads associated with imbalance and stress on the body, mind and conscience. As a result, to a quality-conditioned brain, an income opportunity that pushes the body and mind farther away from homeostatic balance is considered too costly.

Unfortunately, when our *Homo economicus* brains ignore the cost we pay in quality of life (imbalance and trauma) the results could be tragic, as is commonly experienced on the popular paths to fame and fortune. Consider the tragic case of *Glee*, a popular TV musical dramedy that ended in 2015 but became notorious for the fate befallen many of its cast and crew: unexplained sudden deaths, suicides, disappearance, domestic abuses, child porn, drug overdose and divorces. Another example of the cost of trauma (imbalance) is the tragic death of country music star Cady Groves who died of alcohol overdose at the age of 30. She was the third among the siblings to die at a young age following her brothers Kelly and Casey, who both died at 28 (different years) after struggling with prescription drug addiction.

In his book, *When the Body Says No*, Dr. Gabor Mate' methodically explains how ignoring our body's feedback will lead to imbalances and trauma which are associated with numerous diseases and addictions. Yet most of us who are conditioned to run an economic decision-making calculus (as I explained in my own case in the introduction) ignore potential imbalances and trauma that result from our decisions, and their associated costs, particularly in terms of quality of life. To mimic a witty quip from primatologist Robert Sapolsky, the ultimate economic benefit of our quantitative decision-making in life will be landing in an expensive nursing or funeral home!

Reducing Chronic Stress and Pain in Life

Short-term stress drains some of our dopamine. That is why we tend to eat more dopamine-releasing comfort food when stressed. But as we learned, chronic stress, cortisol overload and endorphins released in response to painful stress signals, are all associated with stronger activation of the nucleus accumbens and dopaminergic circuits. As a result, strong or prolonged pain and stress signals may protect us against depression, but make us vulnerable to reward sensitivity and addictions. Homeostasis relies on a healthy natural balance between our pain, stress and pleasure (motivation) responses. We disrupt this delicate balance when we use crutches to bypass our biological feedback loops. Pain, painkillers and even exercise could all become crutches.

Practicing Contentment to Condition our Brain with Serotonin

We learned that serotonin, the *contentment* neuropeptide, keeps us motivated in the face of disappointment - less than expected rewards. Neurologically speaking, The Rolling Stones' popular song *(I Can't Get No) Satisfaction* is an indicator of serotonin deficiency! So we can call this brain chemical the *satisfaction* neurotransmitter too. In fact, research has shown[16] a low ratio of serotonin to dopamine, which leads to impaired serotonergic regulation of dopamine activity, is associated with psychopathy and disinhibition of aggressive impulses.

We can activate our brain's serotonergic circuits by *practicing contentment and simplicity*, an attitude of *"Yes I can, but I won't!"* Religious or spiritual sects like Sufis, Buddhists and Quakers practice *simplicity* as a life principle. Historically, many Quakers, Buddhists and Sufis were wealthy so *simplicity* and *contentment* did not mean starvation or anhedonia. Contentment is a practical and effective strategy for the wealthy and their children to resist getting spoiled (addicted) by dopamine when faced with abundance, novelty and change. *"Yes I can afford that, but I won't buy it!"* For the non-wealthy, contentment saves precious time and resources needed to pursue longer-term, more challenging goals in life.

Perhaps nobody can summarize the real value of contentment better than Jose' Mujica, Uruguay's former poet-farmer-philosopher President, who stepped down voluntarily after one term to attend to his flower gardens: *"Either you're happy with very little, free of all that extra luggage because you have happiness inside, or you don't get anywhere... I am not advocating poverty. I am advocating sobriety... When you buy something, you're not paying money for it. You're paying with the hours of (your) life earning that money. The difference is that life is one thing money can't buy. Life only gets shorter. And it is pitiful to waste one's life and freedom that way."*

Mujica did not wear political masks. A former freedom fighter guerilla and political prisoner-turned-President of Uruguay (2010-2015) he is described as "the world's poorest President." He declined to live in the presidential palace and instead lives on a small farm with his wife. They cultivate and sell chrysanthemums for a living and he donates 90 percent of his monthly salary to charities that benefit poor people and small entrepreneurs because he can live on the income from their sales of flowers.

Practicing contentment and the release of serotonin allows us to appreciate little things in life and become an overall better stronger human by becoming more resilient, empathetic, spiritual, soulful, and strategic in life - remember serotonin is also the strategy hormone. Naturally releasing serotonin is also a preferred way over using SSRI (Selective serotonin reuptake inhibitors) drugs to prevent or fight depression. Surprisingly, SSRIs like Fluvoxamine are now being studied as potential early treatments for COVID-19 or as a treatment for its long term neurological effects (fatigue, foggy brain, etc.). One could expect that the natural release of serotonin may also help against the disease too.

And remember from Chapter 4 that serotonin plays two other major roles. It activates ACC, the brain's center for empathy and soul, and it is released by the action of oxytocin for a healthy response to social and bonding cues (in autism, the oxytocin-serotonin axis is disrupted). So the *contentment* neurotransmitter plays a key role also in our soulful, empathetic and social bonding interactions.

In my household, we use creative ways to practice simplicity, contentment, waste reduction and conservation of natural resources (like electricity, gas and water). As seen in the graph, our energy consumption (where it says You) is about 50% of similar-sized American households (yet still twice the world

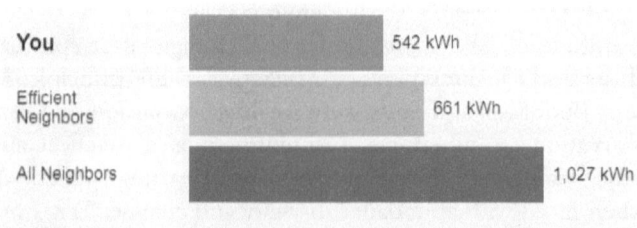

You   542 kWh

Efficient
Neighbors   661 kWh

All Neighbors   1,027 kWh

Jun 15, 2021 - Jul 14, 2021
You're compared with 90 homes that have gas heat. Efficient neighbors are the
most efficient 20% of this group.

average). Our trash (waste) volume is also about half our neighbors'.

I have not always been content or into conservation, but as I got older, contentment (resisting novelty and splurging) has allowed me not to put up with abusive bosses and employers or toxic environments. When we feel trapped or chained, "walking away" is a lot easier if we are already content. Being serotonergic (content) is one of the few ways one can neurologically detach from the addictive dopaminergic allure of money which is used often to keep people in suffocating relationships.

During the early days of the COVID pandemic, I saw on TV and social media, long lines of expensive cars, waiting for hours in food bank lines to receive a $20 donation of food. It was tragic because I knew many of these car owners were chained to long-term car loans that cost them a lot more than the $20 food box they were collecting but they could not walk away from their $300-$600 a month car payment without penalties. I bet you if they knew earlier we could face food shortages or unemployment, and could go back, they would be content with a smaller car and lower monthly car payment. Contentment helps us be resilient.

Being lean and content also places us closer to nature's metabolic efficiency. It makes us more resilient, spiritual, creative and courageous. One who can survive on $1,000 a month can speak and live freer and wear fewer masks than those relying on $10,000 or $100,000 a month.

Contentment and simplicity also allow us to embrace *quality* rather than quantity. We can buy fewer shirts and shoes but better quality ones not made by exploited labor with cheap synthetic materials. So we will encourage producers that focus on quality and good human and environmental relations instead of (cheap) quantity and scale.

I was fortunate to have lived part of my life overseas, raised by parents who experienced the Great Depression and economic recessions and taught us frugality and contentment. I have studied or lived among groups like Sufis, Quakers and Buddhists. I truly believe practicing contentment is an antidote to *Homo economicus* brain's addiction to scalable (boundless) economic rewards. In my opinion, among wealthy nations, contentment is a true form of *free will* in the 21st century. "*Yes I can, but I won't!* And we are not talking about starving or living like Sufis, just scaling down our consumption levels and constant pursuit of novelty.

Contentment is likely to be humanity's best hope to prevent "The Tragedy of Commons," a term used by economists to refer to a situation in which a "common" resource like a meadow, lake, river or airspace ends up being depleted because individual humans, naturally acting in their own economic self-interest, end up depleting the resource for each other and future generations. Self-regulation is a lot more effective than litigation and regulation in curbing excess and greed.

Believing in Higher Purposes and Peace

Rinse Away, Seven Times Over, Stains of Rancor from Your Heart; Only Then, Your Soul Can be Filled, Like a Shiny Glass of Wine With Sweet Intoxicating Love
- Rumi

The term *Eudaimonia* is a Greek word which literally translates to 'good spirit' and is used by psychologists to describe an objective state of unconditional happiness that is achieved through self-actualization and having meaningful higher (non-selfish or materialistic) purposes in one's life. This is in contrast to hedonism. Let us merge the concept with what we learned in this book. In neuroscience, happiness and purpose translate to dopaminergic circuits, which use baselines formed not only by our past behaviors (reward prediction errors) but also by our subjective evaluation of the reward. By assigning high reward values to non-individual, non-economic concepts such as peace we will condition our dopamine baselines. Unlike selfish reward baselines, eudaimonic goals like peace are not depleted (addictive), oversaturated (depressive) or contingent on competition with others, so they never fail to energize us towards happiness. Achieving eudaimonic goals can help release in our brains endocannabinoids, our endogenous 'feel-good' safe opiates and painkillers. Research shows[17] people with mental illness such as depression could often benefit more from endocannabinoids than from medication currently prescribed. This is due to the ability of endocannabinoids to correct the neurochemical imbalances by attacking the source of the issue, rather than just attempting to alleviate the symptoms. Motivating ourselves with unselfish eudaimonic peace-oriented goals makes us high in a healthy way.

Some philosophers and psychologists equate eudaimonia with contentment and spiritual religious feelings. Others define self-actualization in quantitative terms, like climbing higher and higher mountains, literally. There are driven humans who define their eudaimonic purpose and a constant drive to bliss and eternal happiness as the next high mountain peak or personal challenge they conquer. I believe eudaimonic goals cannot be self-serving or scalable (time and quantity). In order for a higher purpose to reverse-condition our economic quantitative brains, it should rewire our brain with the joy of *quality* and *giving* and dissociate it from the concept of *trade*, *quantity* and *time*. Many religions and ancient tribes have practiced rituals of sacrificing valuables (like animals) or making offerings (unconditional kindness) to strangers to condition human brains away from economic calculations based on selfish gains. Spiritual practices remind humans of the benefits of occasionally sacrificing some quantity and convenience.

The following are a sample of spiritual eudaimonic practices:
Buddhism and Zoroastrianism: Buddhism was founded some 2500 years ago in northern India by a wealthy prince who renounced his self-indulgent life to pursue eternal happiness. Buddhists believe in the Noble Eightfold Path which refers to right (pure and unselfish) perspective, right perception, right speech, right behavior, right livelihood, right effort, right mindfulness, and right

concentration (conscious meditation). Followers of Zoroastrianism, founded in Persia around the same time as Buddhism emerged in India, believe in three core principles, shared by Buddhists, of pure (kind and unselfish) thoughts, pure words, and pure behavior.

Sufism and Mysticism: These doctrines are practiced by followers of an ascetic (frugal, non-hedonistic, non-self-indulgent) lifestyle and believers in a common creator and purpose for all beings. Although Sufism is rooted in Islam, its devotional rituals to modesty and unity have influenced Jewish mystics (such as 11th century Bahya ibn Paquda of Spain, followers of Ashkenazi Hasidim of 12th century Germany, and 13th century Abraham Maimonides and Sephardic Jews of Egypt), Christian mystics (such as Meister Eckhart and Friend of God, the 14th century mystic movement within Catholic Church in Germany), and Hindu mystics, yogis and gurus with millions of followers. Later in this chapter, I will share some Sufi stories and parables.

Quakerism and Unitarian Universalism: These are among Christian denominations that promote universal (non-sectarian) values and reconciliation as higher (eudaimonic) causes. Quakers (The Religious Society of Friends) believe in SPICES: Simple life, Peace, Integrity, Community, Equality and Service (Stewardship). Unitarian universalists believe in human dignity, equity, compassion, tolerance, spiritual growth, search for truth, right of conscience, world peace, interdependence with all beings.

Practicing Empathy, Conscience and Consciousness

As discussed in chapter 4, an area adjacent to the prefrontal that plays a major role in goal-directed behavior is the anterior cingulate cortex (ACC) which activates when we experience pain or feel other people's pain (empathy)[18]. Switching off ACC which is also involved in consciousness, self-awareness and free will, leads to perseverative (obsessive, compulsive or addictive) behavior. Contentment, however, is conducive to conscientious behavior as it is mediated by serotonin which also activates ACC.

In addition to breaking addictive feedback loops, pursuing eudaimonic qualitative goals, empathy and consciousness, can also help mitigate our stress responses, cortisol overload and therefore, age-related atrophy (shrinkage) of hippocampus (memory) and cortical (judgment/analysis) regions of the brain. We will discuss conscience more in the context of behaviors that help us detach from our masks, crutches and daggers.

Loving and Trusting

As already discussed, trusting, loving and loyal relationships condition our social brains to release oxytocin which stimulates release of extracellular dopamine within the nucleus accumbens (pleasing and motivating us)[19], which reciprocally activate oxytocinergic neurons, leading to more oxytocin release. This is a self-reinforcing feedback loop very similar to the dopaminergic addictive circuitry but based on the rewards in trust, bonding (parental or romantic) and intimacy so it can help us displace and kick some other addiction cycles. In fact, exogenous (injected) oxytocin is shown to act on Nucleus Accumbens to attenuate formation and expression of tolerance to drugs of abuse. Oxytocin also mitigates the withdrawal symptoms of morphine, alcohol and amphetamines such as Adderall and crystal meth.

Furthermore, as already discussed, oxytocin signals released after monogamous (pair-bonding) mating are received by the (tonic-reward) D2-like dopamine receptors which balance D1-like (phasic) receptors that seem to underlie addictive behaviors and other impulse control disorders. Trust-building seems to benefit pair-bonding species due to its stress reduction effects. Among mammals which are eusocial and pair-bonding, other than humans, naked mole rats are known for their pair-bonding, touchy-feely behavior (they are almost blind so they rely on touching as their main guiding sense) and extreme longevity. Similar to humans, their lifetime number of heartbeats is closer to 2 billion, which is double the lifetime heartbeats in many other species, particularly rodents.

Oxytocin also activates release of serotonin for a healthy response to social bonding cues. New research shows that with autism, the oxytocin-serotonin axis is disrupted.

New research has also revealed that the oxytocin (OT)-secreting system (OSS) plays a pivotal role in the development and functions of our immune system as we age[20]. The OSS can promote the development of thymus and bone marrow, perform immune surveillance, strengthen immune defense, and maintain immune homeostasis.

Among non-pair-bonding species, bonobos seem to be very oxytocin-dependent. Along with the common chimpanzee, the bonobo is the closest extant relative to humans. As the two species are not proficient swimmers, the formation of the Congo River 1.5–2 million years ago possibly led to the speciation of the bonobo. Bonobos live south of the river, and thereby were separated from the ancestors of the common chimpanzee, which live north of the river.

Unlike other great apes, the highest ranking members in a bonobo troop are females. Their matriarchal society uses touch and sex, instead of language, to communicate and resolve all kinds of conflicts. The non-monogamous species seems to properly use the power of oxytocin to avoid testosterone-induced and dopamine-driven violent conflicts common among chimpanzees and other promiscuous primates. Primatologist Frans de Waal has shown bonobos are even capable of empathy and patience.

Many evolutionary biologists today believe monogamous (trusting) relationships have greatly benefited species like humans and gibbons by protecting the kids from infanticide (being killed by their mother's next sexual partner) and by providing children with stable households protected by a vested mother and a vested father. I believe the stabilizing, stress-reducing, addiction-fighting effects of oxytocin and monogamous relations are underappreciated by most researchers.

Stepping Away for Occasional Random Walks

In *The Road Not Taken*, poet Robert Frost writes: *"Two roads diverged in a wood, and I— I took the one less traveled by, And that has made all the difference!"* Frost may not have known it but scientifically speaking, occasional random walks away from well-traveled paths are the stochastic trick used by mathematicians in all sorts of path optimization algorithms[iii]. In other words, if our brains are conditioned by us, or the mobs we follow, to be rewarded only on certain paths, we are not only vulnerable to habituation, addiction and self-delusional behavior, but we will most likely miss some of the best rewards in life that lie outside our rut or loop.

Many successful athletes and business people define discipline as steadfastness on a certain path. But that could also place us on a perseverative path to habituation, and addiction. My definition of beneficial discipline is the ability to drop, *randomly, occasionally and at will,* our rewarding routines in favor of a behavior or activity that balances us. So if we are routinely rewarded by work, exercise, and listening to certain podcasts, our occasional random walk could be a few days in raw nature away from work and the internet, or listening to opposing views and podcasts we usually avoid, or a new exercise routine working a different muscle group, or trying a new food. I remember during the early days of the COVID-19 pandemic, most supermarket shelves were cleared of food items except those of ethnic origin like hummus. I was shocked to see so many people would rather go hungry than try hummus, which is among the most nutrient-packed foods you can find in a supermarket. Random walks that expand our horizons can make us resilient and help us survive. In our family, we strive to maintain discipline, integrity, focus, balance, kindness and flexibility in whatever path we pursue in life.

Beware of the Vicious Triple Hormonal Soup

As explained in chapter 4, the most vicious conditioning cycle happens when our brain and body is cooked by the triple hormonal soup of Testosterone-Dopamine-Cortisol. Testosterone seems to amplify both the reward (dopamine) and stress (cortisol) actions in our brain so it will amplify our vulnerability to addictive pleasure seeking, anxious paranoia or reckless competition. You can identify the action of the triple-hormones in an overly competitive, stressed, hedonistic narcissistic personality, the kind that stops at nothing to attain high alpha positions in dominance hierarchies. Research shows[21] that hormones can get out of whack and stop balancing each other with the struggle for dominance at the very top and bottom of hierarchies. In stable hierarchies of primates like baboons, alpha males experience the highest level of both cortisol (stress) and testosterone (competition) hormones. These hormones usually and naturally balance each other for healthy functioning of the organism. In unstable

---

[iii] We already discussed Levy walks and swarm intelligence. If you are scientifically curious, search online for keywords like "random walk" or "Markov chains" or "Monte Carlo simulations" and meta-heuristic optimization algorithms.

hierarchies, both the very top and bottom of the hierarchy experience hormonal imbalances (high cortisol and testosterone). If hormonal studies in baboons are a guide, many humans must be experiencing the type of struggle other Primates experience at the very top and bottom of dominance hierarchies (high-cortisol high-testosterone, or high-cortisol low-testosterone). This should be humbling for humans who think we are superior and more civilized than primates. We may not be as brutish as primates but are certainly as aggressive, stressed and relentless in our competitions with one another. We just have changed the rules of competition from metabolic (food) and reproductive to economic.

Contrary to testosterone, however, the female sex hormone, estrogen, seems to act as a neuro-protector in times of stress and benefit cognitive tasks such as memory-forming mediated by the hippocampus and frontal lobe.

Drop Masks, Crutches and Daggers

Comedian and social commentator George Carlin once said: *"If honesty were suddenly introduced into American life, the whole system would collapse."*

It's not only America. It's worldwide. We all pay a huge price when we cover up our inner demons and scars with masks. It is hard to ignore the cost we pay for wearing masks when we see the tragic endings for public figures like Michael Jackson, Amy Winehouse, Whitney Houston and her daughter Bobbi, Prince, George Michael, Dolores O'Riordan, Keith Charles Flint, Chris Cornell, Heath Ledger, Brittany Murphy, Paul Walker, Peter Seymour Hoffman, Luke Perry, Carrie Fisher and Lisa Sheridan, just to name a few who seemed to have it all except the ability to remove their masks early on in life. If the road to fame and fortune was a highway, it would be shut down long ago for its casualties and treacherous nature.

And it is not only celebrities who are attached to their masks. A tragic story in recent news was the case of Patrick Rose, the former Boston Police Union President, who reportedly molested the daughter of the man whom he also abused some 25 years earlier (when the victim was a boy). His abusive behavior was masked not only by him but also covered up by colleagues for decades. Were he unmasked earlier his many victims would be spared the lifelong trauma inflicted upon them, their families and people they may have abused in turn. Hurt people hurt people. Masks lead to more masks.

And it is not only humans who pay a cost for using masks and crutches. As already explained, cephalopods, known as intelligent underwater organisms and masters of disguise and crutches, live a short lonely life. In contrast, species like Hyperiid Amphipods resort to *transparency* as a lower (metabolic) cost adaptation strategy than camouflaging (masking). You can literally see through them.

The tragedy of masking is that they lead to self-deception. We may not realize it but before we mask our intentions for others, we always first manipulate *our own* brain because we know other humans, through our collective *conscience*, can sense our intentions. If our brains and psyche were reflective like mirrors, our world would be like those haunted (fun) houses of distorted mirrors. Everyone and every image lies and distorts.

In a world of masks and distortions, having a balanced life is a challenge, particularly for people with public persona such as influencers, celebrities and politicians. What often plagues these diligent, shrewd, thick-skinned and accomplished humans is their struggle to maintain balance in their personal and family lives and relations. While outside home they project strength and equity, at home or in their own body, they are often plagued by imbalance (injustice and disorder). In light of what we learned in this book, true health, peace and justice are achieved not through spending or legislation but through a state of balance in individuals, families and communities that will permeate the society.

So how do we start practicing a maskless life? For starters, we need to stop under- or overestimating ourselves so we are not too fragile or egotistical. One way to dissociate from our masks and veneers (persona) that we have identified with for years is to *deflate* our ego, as Carl Jung suggested. This can result from setbacks in life (such as hitting rock bottom or near death experiences), spiritual or meditative practices (many involve deep breathing and relaxation), contentment (*Yes I can, but I won't*), or use of certain psychedelics that would dissociate and decouple our calculative ego-driven cortical brain from our limbic and subconscious brain. Terence McKenna, the American ethnobotanist who advocated the legalization of natural plant psychedelics believed these plants could help people stuck in ruts to open up to the possibility that what they assumed as truth could be false. I believe there are also non-psychedelic paths to avoid ruts and to reverse-condition our brains, like those paths already discussed as well as practicing conscience and moral values.

Morality and Conscience

This is not a book about religions or spirituality but new research in evolutionary neuroscience confirms many of our traditional moral values have evolved as part of our social conscience to benefit our species.

One aspect of morality is humility and modesty. 13th century Persian poet Saadi narrates the parable of a highly esteemed scholar who had to travel on a sea route. Upon boarding the ship, he asked the seamen if they were schooled in proper grammar and literary sciences. When the unlettered crew pleaded ignorance, the pedant scholar scorned them snobbishly: "I am afraid HALF of your life is amiss, my ignorant friends!" The sailors felt humiliated and heart-broken. Half way through the sea voyage, the ship came across a horrendous typhoon in the dark of night and started sinking. A seaman woke up the scholar and warned him about the impending crisis: "Sir, the ship is sinking, Can you swim?" The scholar said in a grouchy voice: "I am not a seaman. How do you expect me to swim in such waves?" The seaman replied: "I am afraid your ENTIRE life is amiss, my educated friend!"

In America, a strong promoter of conscience, also inspired by Saadi, was Henry David Thoreau. Despite being highly educated, he resisted joining his era's commercial or political mobs and lived a life guided by conscience and empathy. He is known for this statement: *"Can there not be a government in which majorities do not virtually decide right and wrong, but conscience?... Why has every man a conscience, then?"*

Another lifelong advocate of the individual's right to exercise conscience was Thomas Paine, the sociopolitical philosopher and revolutionary mastermind of French and American revolutions. He ended up escaping extremists in both countries when he spoke in favor of personal rights to exercise free will (conscience) and against tyranny of mobs and majority.

In recent times, spiritual yogis like Sadhguru and biologist Bruce Lipton, a pioneer in the field of epigenetics, define conscience as (a spiritual) energy which not only guides our decisions but also conditions our own genome and cells. They believe our hardware (physical cells, genes and body) are merely vessels reflecting our (spiritual) conscience.

Other scholars approach conscience from an evolutionary metabolic perspective. In her book, *Conscience: The Origins of Moral Intuition*, neurophilosopher Patricia Churchland discusses how the human brain as a social brain is wired and adapted for social conscience (common knowledge of moral norms). She argues that compassion, kindness and other morally intuitive traits are socially adaptive norms which evolved as a sociobiological solution to reduce constraint satisfaction costs for each ecosystem. In other words, without moral traits like mutual altruism and compassion there would be huge social and metabolic (biological budgeting) costs associated with conflict resolution and stability of human ecosystems (societies).

In fact, new research in moral psychology and neuroscience supports the *Grace hypothesis* which asserts that human brains are metabolically evolved for honesty so dishonest behavior (and wearing masks) is metabolically tolling (on our brain), hence maladaptive (on a group level). When study subjects (undergoing functional magnetic resonance imaging, fMRI) had a chance to be dishonest for personal gains, individuals who behaved dishonestly exhibited increased activity in control-related regions of prefrontal cortex, both when choosing to behave dishonestly and on occasions when they refrained from dishonesty[22]. In contrast, individuals who behaved honestly exhibited no additional control-related activity (or other kind of activity) when choosing to behave honestly, as compared with a control condition in which there was no opportunity for dishonest gain. Levels of activity in control-related regions of the prefrontal cortex correlate with the frequency of dishonesty in individuals. In other words, we become more metabolically efficient and homeostatically healthier the less we practice dishonesty and the fewer masks we wear. So we can say, for humans, morality and conscience are evolutionarily beneficial cognitive adaptations.

One of the control-related regions of the prefrontal cortex, which plays a major role in goal-directed behavior is the anterior cingulate cortex (ACC). We already discussed how ACC is involved in empathy[iv] (feeling other people's pain) as well as our own consciousness, self-awareness and free will. When ACC is switched off, we resort to perseverative (obsessive, compulsive or addictive) behavior. The fact that our own self-awareness, free will, pain sensation and cognitive strength overlaps with our social empathy in our brain, can be used as another proof that the cognitive cortical human brain has evolved around sociality and group selection. We are most self-aware and least prone to addictive and obsessive behavior when we activate the empathy-driven, morality and conscience-oriented parts of our brain. Research on psychopathic individuals or powerful narcissistic people[23] has shown they have decreased activity in the anterior cingulate cortex (ACC) during decision-making. Since ACC is at the intersection of human cognition and emotions, I call it the cerebral seat of our *soul*. It is probably what makes us unique as compared to other self-aware intelligent species. Otherwise, when considering awareness in terms of brain surface, humans are in second place since porpoise and dolphin brains have a larger surface area than human brains.

Collective consciousness, or *Noosphere*, is a utopian concept first proposed by biogeochemist Vladimir Vernadsky, and philosopher and Jesuit priest Pierre Teilhard de Chardinrecent. As mentioned earlier, a noosphere is a thought sphere that encompasses our collective global human consciousness, and is the evolutionary equivalent of a biosphere as a human-contributed layer on the planet. Research shows it is immensely beneficial to humans when they harmonize their thoughts, as in spiritual group activities, or their actions together. An example is "group drumming" which is shown to significantly improve depression, anxiety and social resilience among users compared with a non-music control group[24]. It has also been shown that humans can naturally synchronize their body functions as in the case of synchronization of menstrual cycles occasionally observed among women who live together.

Feedback 24/7: Biomarkers of Balance

Our species is plagued by feedback loops that are self-reinforcing, broken or shunted by our masks and crutches. Although our brains are connected to our metabolic bodies, by using all sorts of biohacking tools (medicine, supplements, artificial lights, machines, etc.) we are helping our brains evolve to further ignore our body's feedback. It is like playing a video game, in which we can fly, run and fight countless opponents without dying because we can respawn and have unlimited health (energy). So it is in our own self-interest to start dropping masks, crutches and daggers, to develop a thick skin, and encourage, exchange and look for honest real-time feedback anywhere we can, and to watch for all the triggers of a self-delusional brain outlined in chapters 5 and 6.

The world around us, our looks and skin appearance, mood, temper, blood biomarkers, our families, friends, relations, workplaces, societies, politics,

---

[iv] Other areas of prefrontal cortex involved in cognitive control are dorsolateral prefrontal cortex (DLPFC), and the ventrolateral prefrontal cortex (VLFPC).

healthcare and political systems all reflect our choices and imbalances if we pay close attention. The world is nothing but an echo chamber, as stated in the opening poem by Rumi. For a self-delusional species with numerous blind spots and addictive personality, feedback is the best gift. We should stay away from people with daggers at all costs. Instead, it is adaptive, both personally and evolutionarily, to surround ourselves with true *maskless* friends who give and take honest feedback. Instead of joining herds and like-minded folks, it will serve us to hang out with people, or follow influencers, who are more balanced or healthier than us or help us become more balanced in life. We should always use balance as a selection guideline in all social relations (friendships, love and work relations). Think about you on one side of a seesaw, grossly imbalanced. You want someone to balance you on the other side. Ideally, you want you and your balancing partner to be light and closer to the center (pivot). Too heavily imbalanced sides on a seesaw will break it instead of balancing it the same way allostatic loads on our body will break us down over time (as seen in chapter 5).

To develop a mindset striving for balance, we need to reward and seek feedback. As wisely stated by Nietzsche *"The strength of a person's spirit is measured by how much 'truth' he could tolerate, or more precisely, to what extent he needs to have it diluted, disguised, sweetened, muted, falsified."* We should all practice selfless acts of honest feedback and communication like what parents of Cody Holland did, He died at the age of 25 in Ogdensburg, New York[25] but his parents took the pain to deviate from traditional formalities and include some bitter sweet truth in his public obituary to benefit other humans:

"He loved with his whole heart. Cody loved his family, both biological and those he selected to be his family. He loved his black wife-beater t-shirts, his trucks, his motorcycle, his dog, his girl, his guns, huntin' and fishin' with his friends and family. Not necessarily in that order... He loved his middle finger and showing his butt to the world! In his younger days, he was pretty sure he was a gangsta! Gold chains, hat on sideways and pants down to his ankles. Some things don't change. He grew up, but still loved inappropriate t-shirts, ball caps, and big belt buckles.. When asked about future endeavors, Cody's response was, 'I'm gonna be a porn star!' However, he did get a degree from Paul Smith's and became a NYS Corrections Officer...Cody James left us on August 8, 2020, as a result of injuries sustained from being a dumb ass. He drank, drove, and didn't wear his seat belt! Please, don't be a dumb ass!.. Please come to the calling hours dressed how you are most comfortable. This is how Cody would want it"

The truth often hurts but it shall ultimately set humans free, at least metabolically, as we saw with the cortical metabolic load on our brains when we try to hide or sugarcoat stuff. Suicide rates seem to be higher in societies like Japan in which people do not exchange direct and free flowing social and personal feedback.

To ensure that your feedback is sincere, ask yourself if it is more about the person's blind spots or your need to vent out? And be prepared to receive feedback when you dish it out. We all need better feedback loops from each other, not sermons. In addition to exchanging truthful feedback with each other, we need to pay close attention to our own body's signals. As a self-delusional tool-dependent species, we can easily ignore, misinterpret, or patch up, using crutches, the many warning signals our own body issues on a regular basis. Our brains, being in the driver seat of modern human's bodies, can easily self-deceive us.

Generally speaking, the world around us is a mirror to our imbalances but the following are some of the specific feedback signals we should not ignore. These are important individualized signals because each person is unique in his or her response to external or internal stimuli or even to medical interventions. This is not a book about physiology so after reviewing this chapter, you can consult about the following whole body biomarkers with integrative or functional medicine practitioners (or check my blogs for reference to such organizations):

1) Hormonal biomarkers: Blood biomarkers and ratios of testosterone, cortisol, estrogen thyroid, and if possible hormonal metabolites in the urine (once a year). Remember many hormones have a diurnal pattern so staying healthy is like walking a tightrope of hormonal balance.

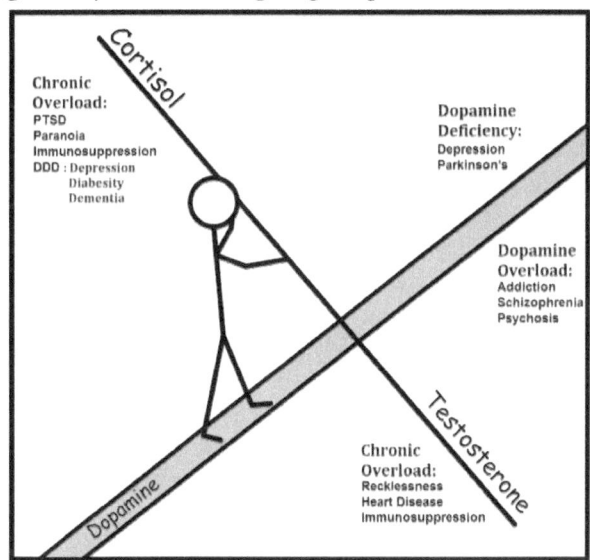

2) Blood test, once a year: Complete blood count (CBC) including concentration of red blood cells (for blood viscosity), white blood cell types (the ratio of neutrophils to lymphocytes for example, as a measure of load on your innate and adaptive immune system), alkalinity of blood, Vitamin D level (30-60 ng/mL seems to be ideal), and lipid profile, For example ratio of Total cholesterol to HDL (less than 3 is ideal) or Triglyceride to HDL ratio (1 is close to optimal)

3) Regular monitoring of resting heart beat (lower is usually better), fasting blood glucose, blood glucose 1.5-2 hours after eating, and blood

pressure ranges when exposed to different stressors (a measure of activity of the sympathetic nervous system).

4) Whole body feedback signals: As already mentioned, unlike reductionist medicine which relies mostly on organ level evaluations, ancient, holistic, integrated and functional medicine rely heavily on organismic (whole body) feedback signals to evaluate someone's overall state of homeostasis. As described in Chapter 2, this is similar to the system's approach we use as engineers and scientists to model and control complex systems towards steady state. The focus is applying conservation laws to the input and output of the whole system and monitoring dependent variables. Except in the human body, instead of sensors used in industrial process control, we have bodily signs and symptoms like red or pale skin, baggy eyes, constipation or dry stool, etc. In the human body, there are many signals that correlate well with the body's overall level of homeostasis.. For example, it is now evident that a strong craving for chewing ice (pagophagia) is associated with iron deficiency, with or without anemia. Our other cravings can also be linked to our mineral or hormonal imbalances. The main telltale signs of our overall balance include: Our sleep patterns (the time, ease, depth and duration of sleep), stool habits and excrement (frequency, morphology, color, dryness, density)[v], urine characteristics (frequency, volume, color, turbidity, pH[vi], etc.), other efferent body fluids (like smell and intensity of sweat or runny nose), our weight fluctuations on a daily basis, our appetite and cravings, our pains (location, intensity, frequency and shifting direction), our skin elasticity and turgor, especially under chins, under eyes, on our neck and dorsal (back) side of our hands, our body proportions, particularly ratio of belly (visceral fat) to legs and upper body, our energy to play in nature (as opposed to work indoors), and obviously healthy libido, intimate relations and sexual satisfaction, which as described earlier, is almost a perfect biomarker for a balanced sympathetic and parasympathetic response in humans. The physiology of erection and relaxation of smooth muscles seems to be mediated by nitric oxide[26], the body's all-important messenger molecule for vasodilation (expanding of blood vessels to lower blood pressure), which is also linked to cellular growth, angiogenesis, insulin metabolism, immune response and cardiovascular health. That is why some physiologists correlate erectile health with cardiovascular health. Chronic stress and sympathetic response depletes

---

[v] Constipation and disorders in stool habits and morphology are important early signals (feedback) for many diseases, including kidney stones

[vi] I use a pH meter on a regular basis to ensure through proper diet and lifestyle I avoid the oxidative stress and acidic pH range (below 6.5) that trigger my kidney stones.

acetylcholine, the neurotransmitter that signals vascular cells to produce nitric oxide[vii].

Recent studies have provided evidence[27] for the neuromodulatory effects of serotonergic, noradrenergic, and dopaminergic agents on neural substrates of erotic stimulus processing. In other words, a healthy balance of sympathetic and parasympathetic hormonal axes ensures proper functioning of the motivational, emotional, autonomic, and attentional components of the sexual response.

There are other body signals that correlate with our stress response. For example, a group of researchers have correlated people's skin properties to their level of anxiety and addiction to the internet[28]. They have shown that our skin conductance response (SCR) is an indirect measure of our sympathetic autonomic activity that is associated with both emotion and attention. Stressful or painful stimuli (even pin pricks) elicit a sympathetic response by the sweat glands, increasing secretion. Although this increase is generally small, sweat contains water and electrolytes, which increase electrical conductivity, thus lowering the electrical resistance of the skin. Another common manifestation of galvanic skin response (GSR) is the vasodilation (dilation) of blood vessels in the face, referred to as blushing, as well as increased sweating that occurs when one is embarrassed.

I am currently researching a list of important body signals that scientifically correlate with our homeostasis and health. In my blogs I will explain in detail how anyone can easily monitor these body signals as a measure of overall health and to control/correlate them with lifestyle and diet choices. I also regularly post what I find in research papers.

Drop Crutches

We defined crutches as tools that make us dependent and even addicted. By that definition, if we have become dependent on spending money to buy happiness (medicine, drugs, alcohol, tobacco, vacations, apparel, shoes, entertainment, gadgets, cars, games, toys and tools) or on something we *really* don't *need*, we are using money, not only as a tool, but as a crutch. We can practice reducing our dependence on spending money for non-essential items by spending more time (less money) on inexpensive interests. Remember, the experiential value of time is reflected more in the quality of the outcome than its quantity. Think about the quality of nutrients in fresh fish you catch in a clean river (unfortunately no longer easy to find), the bread, yogurt, kefir or sauerkraut you ferment at home, or the nuts you extract by cracking open fresh hulled walnuts, pecans or pistachio. The more time we spend to treat ourselves to quality, the more we condition our brain to wean off seeking pleasure in expensive habits and formatting itself in quantitative economic norms.

Reliance on crutches makes humans easier to control, as envisioned by clairvoyant authors like George Orwell and Aldous Huxley. In Orwell's *1984* (published 1949), people were controlled by pain and fear (the stress and fear-

---

[vii] Global sperm counts and fertility levels are expected to drastically drop within the next three decades.

induced self-reinforcing cortisol circuits) and books were banned. In *Brave New World* (published 1932) Huxley foresees a dystopian world in which humans are not conditioned by fear but by individual hedonistic sexual pleasures (dopaminergic circuits). Books and activities in nature are discouraged and people are addicted (crutched) by free government-provided pills (called *soma*) to manage or avoid pain, depression and mental stress.

Towards a Balanced Socioeconomic System

As discussed in chapter 7, genuine education, peace and health often become orphans (no sponsors because time consuming and not profitable) in scalable economies which rely on spending money to manage disease, disorder and discord. So like a chemical reaction which automatically moves towards chaos (higher entropy, lower Gibbs free energy as discussed in chapter 2) our societies will continue creating chaos, disorder and disease unless we all strive for balance and equilibrium, as nature and life does. Without balance as our goal in every personal or social decision, we will always remain on the edge regardless of how rich or economically productive we are. Many of us will be one disease away from disability or death, one paycheck away from financial crisis. Our societies will be one economic (like supply chain) or biologic (like pandemic or biological agents) crisis away from pandemonium and paralysis. Here are some thoughts on how balanced-driven systems could look like.

On Our Economic Brains

I recently had a window into the economically-driven brain of a modern human. A middle-aged woman interviewed on our local TV station was visibly outraged about her workplace being closed due to the COVID-19 pandemic. She was not upset about an income loss because she received generous unemployment benefits. She shared her frustration with the cameraman: "You cannot expect me to stay home all day with my children. I need to be productive again." She was thinking exactly like a *Homo economicus* whose brain is conditioned by economic productivity in contrast to the 4F's of natural evolution discussed in chapter 4. In nature, work (foraging and hunting) is to support sustenance, reproduction and relaxation not the other way around.

My other recent exposure to the brain of a *Homo economicus* came when I saw a callous tweet by a slick-looking salesperson mocking people who debated mandates and pandemic policies: "While you were arguing if the glass is half full or half empty, I sold it!"

My third personal observation of how our economic brains work relates to lawn mowers. Today, most suburban American households with a lawn or yard own their own riding tractor lawn mower, typically costing about $2500 each and used for about four months in a year and on average about one hour every 10 days, for a total of 10 hours a year. To park their tractors, many people install a tool shed, costing typically an additional $2500. So a typical suburban American neighborhood consisting of 20 homes invests about $100,000 for the convenience of riding while mowing (about 10 hours a year). Many of the same suburban households spend about $100 a month on health club or gym memberships, plus at least 12 hours a month in gyms or commuting to gyms, to stay in shape. This type of resource-utilization would never be allowed (adaptive) in a metabolically-efficient balanced natural ecosystem. Why? For starters, let's translate the situation into labor and energy equivalents. $5000 spent on a tractor and shed is about one month of after-tax income or labor productivity of an average household or about 200 hours in labor and commute-to-work. 12 hours a month in or on the way to the gym means another 240 hours a year. One could use a small inexpensive push mower and about 4 hours a month (⅓ of that 12 hours) to burn enough calories and stay in shape both physically and financially without the need for the tractor, shed and gym membership. Imagine what higher causes people could serve with that $100,000 saved. That's your college tuition for one student in each neighborhood. Why don't more people use push mowers? Because our brains, decisions and lifestyle are now driven by economic productivity (quantity, convenience, speed) and not individual metabolic efficiency and balance! One drives spending and quantity, another drives frugality and conservation. One expands GDP, another shrinks it (and our spending).

Our scalable economic systems rely on the unlimited scalability of production mostly in response to unlimited scalability of our problems. Humans have replaced natural feedback loops with feeder systems and cyclical economies. Organizations such as the Center for the Advancement of the Steady State Economy (CASSE) are now advocating for economic policies and goals which, like natural systems, achieve steady states as a result and not unlimited growth of GDP.

Throughout this book, I have used science-based examples from nature to demonstrate that human-specific socioeconomic problems (imbalances and disorders) cannot be fixed with money alone until we reach a naturally balanced state of metabolic efficiency and homeostasis (steady state) at an individual level. I explained how injustice in wild nature is nothing but imbalance which often corrects itself with time. As modern humans, however, we have built our economic system on collective scalability and not individual balance. Many people are now picking on companies like Amazon and Walmart as exploiters-in-chief without realizing that the economic system rewards *any* company that helps humans scale or speed up at the lowest cost. Bankers make it easy for people to buy larger homes they could barely afford (scale up their budget). Pharmaceutical companies help us with a speedy recovery from pain and diseases of imbalance (and scale up our productive hours without healing). Airlines allow us to travel farther and faster (scale up our reach and economic productivity). Cell phones allow us to live farther from our customers and even from people we love (scale up our reach).

To curb the excesses, imbalances and injustices associated with scalable economies, we need to *individually* become more balanced by setting limits to our quest for speed, quantity, convenience and realize, as we learned in Chapter 3, that nature is always about trading off quality with quantity and resilience with speed. This may be why some traditional sects such as the Amish do not use phones because they believe family, friendship and community ties are only possible in face-to-face personal meetings. Members of the Amish community often live close to each other and meet at least once a week with their friends, parents, children, grandparents, grandchildren and neighbors. Their community is physically close-knit (local, not virtual or remote) that is deeply involved with supporting individual families. Compared to a *Homo economicus* brain that assigns high reward values to quantity, speed, reach and convenience, the Amish brain probably assigns a higher reward (utility) value to unit time spent on traditional (*Homo sapiens*) life experiences such as regularly meeting friends and family face to face. My father lived and thought very much like the Amish but I lived and thought like a *Homo economicus* for a good part of my life.

The problem with *The Sky's the Limit* mindset is that many humans now even view their lives as scalable quantities, like their income and stock prices, which they can extend forever. Death avoidance at any cost is actually costing the world a lot, particularly in wealthy nations and among the ruling class. Longevity research and medical and genetic treatments for aged populations are among the most lucrative scientific fields. I already cited the new Alzheimer's drug costing $56000 per person per year. There are many others on the way.

Stories and Parables from Sufis and Mystics

Interestingly, in some eastern mystic traditions, the fear of death is associated with attachment to the material world. Legend has it that Attar of Nishapur, the 12th century mystic Persian poet and an inspiration to Rumi, ran a flourishing perfume shop in a bazaar (*Attar* means perfume merchant) until one day a wandering dervish entered his shop and asked for alms. When Attar refused, the dervish asked: "*If with all your wealth you can't part with a few coins how will your soul part with your body when the day comes?*" Attar replied: "*I will die exactly like you do, reluctantly!*" The dervish replied: "*You sure?*" Then he paired his galoshes (simple flat shoes from Persia) on the floor as his pillow, lay down, invoked the Divine and died right there and then. Attar was so shocked and moved by this act of devotion that he closed his shop and began his spiritual journey. Rumi refers to Attar as "*one who trekked through all the seven stages of selfless devotion and love.*"

Sufis (followers of mysticism or *Irfan*) do not *learn* from teachers and lecturers but *follow* gurus whose words and actions are the same and effortless. By living in harmony with a unique universal conscience, they strive not to upset the balance that connects all life forms and materials.

Sufi masters did not lecture and were known for brevity when replying to questions by followers. There are a good number of spiritual and inspirational stories and quotes about Sufi followers which are not published in the West because the sources are in Persian, Arabic or Turkish. I include some of these stories in my blogs and writings. Here are some examples.

A story quoted by Attar relates to a Sufi guru called Hassan of Basra (Basri) who was asked by one of his wealthy followers: *"O'Shaikh, What's the worst fate and punishment in this world?"* Basri replies: *"The death of our spirit and soul."* Follower: *"How would that happen?"* Basri: *"When we get attached to our assets."*

Attar quotes another story about a sufi named Oveis of Gharan (an old town in Yemen) or Gharani, who was known for solitude and meditation. Once he was asked by a famous spiritual guru (an influencer, in today's jargon): *"Whom should I befriend and how can I reach more followers?"* Gharani replied: *"The one truly touched by the spirit, would neither seek more followers nor more friends."*

Attar quotes another Sufi, Mohammad Vase', when he was asked by a wealthy follower: *"How to live like a king but without the worries and guilt?"* Vase' replies: *"Become content, then you shall feel needless like a king, but without his worries or sins."*

Vase' is also quoted for this observation: *"You can't call a person shrewd whose number of remaining days on this planet keeps shrinking while he keeps expanding his number of headaches, assets and regrets!"*

On Our Politics and Society

To turn politics away from its current budgeting and money-printing, money-spending role, we need to value and reward, as individuals and therefore as a society, *balance* in public affairs. In other words, we need to reward public servants as problem solvers that balance budgets while *reducing* levels of disorder, disease and discord. Public servants' term extension or salary/pension would be linked to unpacking our hospitals, courthouses, prisons while shrinking the collective social expense (budget deficit) on courts, police, hospitals, army, prisons, etc. In a balance-oriented system, the less disorder and imbalance we have the higher we would reward the problem solvers among us. This would attract more balance-oriented, unselfish, strategic problem solvers to public service. We all realize that any social system is composed of people, processes and products (3 Ps), yet as a self-delusional species, we expect healthy outcomes from severely tainted or defective (imbalanced) processes, people and products!

We can promote a healthier society by rewarding those among us who do their part to be healthy, balanced and peaceful, and parents who raise such children. As described in chapters 1 and 7, our current economic system spends huge sums on containing dysfunctions and disorder, in prisons, schools, and hospitals. Our current budgets and incentives are mainly towards containment and not prevention. I strongly believe that if we financially rewarded parents for improving their children's health and school performance in science, reading and writing, we would see drastic changes in the society and parents' engagement. We need to retool our socio-political and socioeconomic systems for achieving balance and conservation of resources (natural and human) as opposed to scalable growth and spending.

A society is a network of individuals so its balance is a dependent variable and a function of balance of individual members and their connecting links (relations). That is why it's important to have balanced, healthy smart individuals in our relationship networks. Research shows that the social clusters we belong to have a contagious biasing effect on our behavior and traits, especially if the connection to the cluster is rigid or strong. In other words, if most people in our network of friends are smokers or spenders, we will also be biased towards smoking or spending even if our cognitive brain directs us otherwise. That is why it is wise to surround ourselves with people who are more balanced, wiser and healthier than we are, or at least don't have the same biases we have. That is why for critical decisions in complex matters, besides doing my own research, I seek advice from a few people in my network who are not only well-read on the topic but are also not self-censored or biased (see the list of biasing factors in Chapter 6). For example, on the risks and benefits of certain medical treatments or drugs, I will seek advice from chemist and biologist friends but not those whose employment with pharmaceutical companies or hospitals may bias their views.

On the social level, a balance-oriented system will reward social feedback and transparency on all levels. We would encourage and protect the whistleblowers among us, not punish them. We would seek feedback from outliers among us not censor or ostracize them. We saw how historically good independent science originated from outliers not the authoritarian ones who pushed radical mastectomies, defective drugs and vaccines, and lobotomies. We should minimize secrets and maximize transparency on all aspects of society. The more closed doors, secrets, and privileged access a society has, the farther it gets from a state of self-correction, balance and equilibrium.

Many of us are now taught to expect miraculous solutions to our problems from politicians and technocrats. Pointing fingers at politicians and billionaires may be popular and justified but it has become a crutch which is *not* solving our problems. Socioeconomic problems are rooted in our metabolic imbalances and need *long-term, bottom-up, human-centered, metabolic and balance-oriented* solutions. Politicians and CEOs, by the nature of their job and personalities, are into *short-term, top-down, technocratic, economic or bureaucratic* approaches. We want them to educate, feed and protect us. We demand from them quick fixes and they deliver to us quick fixes, i.e., crutches and interim remedies such as printing more money (which is practically borrowing against future) or reprioritizing budget outlays. So we kick the can to them and they basically kick the can down the road. And when funds run out of crutches and Band-Aids don't work, we demand change and vote for the next anti-establishment candidate for change. The cycle of popular outrage and political appeasement continues as we ignore the fundamental natural imbalances that plague our bodies, minds and communities.

When it comes to social justice, and the astronomical cost and inefficiency of our legal system (chapters 1 and 7), I am reminded of Voltaire's statement about the roots of inequity and injustice[29] which is aligned with some of the concepts shared in this chapter, including contentment, moderation, scaling down and metabolic efficiency. Voltaire believed to the extent humans enjoy natural faculties or perform natural functions, they cannot be treated unequally or exploited so injustice is rooted in our needs and dependence on a human-made economic and legal system:

"The King of China, the Great Mogul, the Padisha of Turkey, cannot say to the least of men: 'I forbid you to digest, to go to the privy and to think!'.. Animals by nature have over us the advantage of independence. If a bull which is wooing a heifer is driven away with blows of the horns by a stronger bull, it goes in search of another mistress in another field, and lives free...If this globe were covered with wholesome fruits; if the air, which should contribute to our life, gave us no diseases and a premature death; if man had no need of lodging and bed other than those of the buck and the deer; then the Gengis-kans and the Tamerlans would have no servants other than their children.. In the natural state enjoyed by all untamed quadrupeds, birds and reptiles, man would be as happy as they; domination would then be a chimera, an absurdity of which no one would think; for why seek servants when you have no need of their service?.. If it came into the head of some individual of tyrannous mind and brawny arm to enslave a neighbor less strong than he, the thing would be impossible; the oppressed would be on the Danube before the oppressor had taken his measures on the Volga! All men would then be necessarily equal, if they were without needs; the poverty connected with our species subordinates one man to another: it is not the inequality which is the real misfortune, it is the dependence."

On Our Ecosystem

We learned how nature often works through symbiosis, equilibria and conservation to maximize the Gibbs free energy. We can minimize our disruptions to that rhythm by individually being metabolically efficient and also by not adding concentrated human-made systems to the ecosystem, like industrial farms. Small scattered family-owned farms that use manual labor, basic technologies and no harmful chemicals, would have a small impact on the ecosystem's overall equilibrium and health. But large human or livestock population centers are different. By some estimates[30], humans and our livestock now make up 97 percent of all animal biomass on land. Wild animals (mammals and birds) have been reduced to a mere 3 percent, some 75% less than the start of farming age. A research paper reveals[31] the historical impact of humans on the global biomass of prominent taxa, including mammals, fish, and plants. For instance, the global marine biomass pyramid now contains more consumers than producers, thus forming an inverse food pyramid not common in natural evolution. On the other hand, the mass of humans is now an order of magnitude higher than that of all wild mammals combined.

It is not only the natural ecosystems we impact. Our metabolically-inefficient, imbalanced scalable hierarchies have made us experts at creating top heavy, inefficient, human-made (artificial) ecosystems. Just consider our universities and bureaucracies as prime examples of how we create large overheads in systems we build. Economically-speaking, the concept of trickle-down (tax cut for the rich) to prevent a top-heavy income distribution has not worked. Research by economists David Hope and Julian Limberg in 18 developed free-market countries shows that income disparity has grown over time with tax reduction for the wealthy. But the policies of welfare for the poor have also proved to be inefficient and wasteful. The problem, as stated earlier, is an economic system that rewards scale, speed and quantity as opposed to balance, metabolic efficiency (conservation), quality and resilience.

There are those who blame technology for large-scale environmental damages. Along with his vital work as a bacteriologist, Rene' Dubos was also a pioneering environmentalist. In an interview cited in a 1982 New York Times article, he said:

"I have been reading predictions of the future... In all cases, the future of which they speak is merely a grotesque extension of the present – simply more and more loading of our environment with the waste products of technological civilisation. In my opinion, there is no chance of solving the problem of pollution – or the other threats to human life – if we accept the idea that technology is to rule our future."

Dubos may have a point which is shared in Dr. Seuss' 1971 book *Lorax* and the brilliant 2012 animated movie based on the book, foreshadowing a world in which technological advances lead to sales of fresh-air and exploitation of nature for profit. When I recently saw the black market and long lines in some countries for oxygen tanks, or the shortage of bottled water in some US stores, I could not help but think about *Lorax*! It's a worthwhile movie for children and adults to watch and discuss.

On Our Education and Science

Our schools which promote individual competitiveness fail to note the benefits of swarm or collective intelligence, as seen in other eusocial species or even in flocks of birds. Even in humans, researchers have found evidence that a group's performance on a wide variety of tasks depends on a general *collective intelligence* factor which is not strongly correlated with the average or maximum individual intelligence of group members, but with the group's social cohesion and with the equality in distribution of conversational turn-taking[32]. In other words, when every member of the society contributes to the dialogue, there is a synergistic effect on the collective intelligence. A super competitive educational selection process turns out human equivalents of Super chickens, among them psychopaths.

Yet as tools and rules replace natural fitness (balance and metabolic efficiency) to become our new keys to adaptation, our toolmakers - specialists, doctors, scientists, coders and engineers - and rule makers - lawmakers, lawyers, bankers, politicians, religious leaders, judges - become the alpha males and females in the human hierarchy. Many other humans who serve the toolmakers and rule makers live on the fringes of society, one paycheck away from hunger and poverty, one pill or surgery away from death. The following is a good example of how tools made residents of an ancient civilization unfit and dependent.

Unlike most countries on the planet, Bhutan follows not GDP but GDH, Gross Domestic Happiness to guide its social and economic policies. GDH is an index which measures the collective happiness and well-being of the population. GDH does include scalable quantitative measures like per capita income but also includes balance of work and life and sleep, ecological biodiversity, vitality and safety of communities and families, mental health and emotional balance, cultural diversity and resilience. Yet even the country of Bhutan, known for its happy healthy residents, has gone through major changes in recent years. Bhutan was among the world's last technology holdouts. But pressured by international organizations like the World Bank, the country expanded its electricity network from 0% to 90% of households between 2000 and 2018. During the same period, the incidence rate of diabetes went from 3% to 12% of the population, without any major change in the country's diet. The alarming trend could have resulted from the impact of electromagnetic fields, the use of refrigerator and therefore process and non-fresh food instead of their traditional fermented and fresh food, the use of rice cookers as opposed to traditional cooking which involved rinsing starch, and obviously the use of TVs, computers and mobile phones leading to a more sedentary lifestyle.

Modern reductionist science and technology provides us with amazing tools to increase our productivity and convenience but it often trades off our natural resilience. Take the field of immunology. Our naturally evolved immune system, when balanced and healthy, has many layers of defenses against pathogens: First line: Skin, mucus membranes (IgA), mast cells, macrophages, Langerhans cells, Interferons; Second Line: Inflammation, natural killer cells, cytokine messengers, diapedesis, fever, vasodilation, phagocytosis; Third line: B -Cells (Specify, identify, activate, release antibodies in plasma through Monoclonal replication and somatic hypermutation); and Fourth line: T Cells (cell-mediated, Helper for B-cells and Killer for self-antigens or damaged cells). Disease is a manifestation of an imbalanced overactive (like auto-immune diseases) or underactive (cancer and type-II diabetes) immune system.

The proper activation, evolution and balance of the immune system require time because each person has a different and unique immune system. Generally-speaking, women and younger populations have a stronger humoral immune defense (like B cells) than older male populations which have shifted to a more cellular (T cell) immune strategy. This makes women and young people more prone to auto-immune risks (with an overactive immune system) and older males more prone to cancer and viral infections. One-size-fits-all solutions are therefore not optimal and could result in more adverse reactions in women and young men[33].

Yet, in modern times, we cannot wait for weeks to heal or risk getting sick and bed-ridden, so we use all sorts of medical interventions that will evade, fool, tweak and override our own immune system before using the same system that it outsmarted (upset) to fight disease (imbalance). Needless to say, such interventions disrupt the natural progression of our own innate and adaptive immune system processes.

In this process, we use medicine for a speedy victory of individuals over certain strains of bacteria or viruses at the expense of collective long-term resilience of species against the spectrum of pathogenic bacterial or viral strains. We have heard about the super-germs now demonstrating antibiotic resistance, or leaky vaccines with nonspecific (immunosuppression) effects. We have also heard about auto-immune diseases, cancer and viral infections, all hallmarks of imbalanced immune systems. The intrusion of oil-based chemicals and plastics, synthetic, denatured and artificial food ingredients into our body further confuses our body's naturally-evolved immune system and its detection of foreign agents. And the more imbalanced (diseased) we become, the more we need to rely on additional interventions and crutches as quick fixes.

As a metabolically inefficient, imbalanced and impatient species, we demand quick fixes from our toolmakers and rule makers and they happily oblige. In scalable economic models, efficient free markets are those which match supplies to demand and encourage demand. But more demand means more problems and disorders. So when our problems persist or confound, or when new imbalances arise (such as side effects of medicines or new technology), we blame the industry and ask for new quick fixes for the side effects and imbalances originally created by quick fixes we asked for. Remember *Crutches Beget Crutches*. Many people do not even know who to trust anymore. Others blame corporate or government-funded scientists and technologists for distorting the truth. In a 2015 paper, Carlton Gyles, laments on how the relationship between pharmaceutical companies and academic physicians at prestigious universities impacts certain drug-related publications and the marketing of prescription drugs:

"Potential conflicts of interest seemed to abound: millions of dollars in consulting and speaking fees to physicians who promoted specific drugs, public research dollars being used by a researcher to test a drug owned by a company in which the researcher held millions of dollars in shares, failure of university researchers to disclose income from drug companies, company subsidies to physician continuing education, publishing practice guidelines involving drugs in which the authors have a financial interest, using influential physicians to promote drugs for unapproved uses, bias in favor of a product coming from failure to publish negative results and repeated publication of positive results in different forms."

These are tragic and valid observations about shortcomings of systems that have delivered us products we demand. But at the root of all this, as I have indicated throughout this book, lies our departure as a species from a natural evolutionary path.

In July of 2020 humans successfully landed NASA's Perseverance Rover on the surface of Mars some 471 million kilometers away from us. The rover has 23 cameras, including some that can take selfies, UV and Laser Microimagers, X Ray spectrometers and other amazing gadgets to detect even fossilized microbial life. As a former NASA grantee, I was thrilled to follow and hear the news. Yet I was still bemused about comments made by President Trump some seven months earlier about an earthly mundane problem impacting him and Americans, the nation that sent Perseverance to Mars: The number of times toilets have to be flushed! Trump had a point. Many new eco-efficient toilets, conserved water but at the expense of lower flushing and suction volume, which meant more frequent flushes for a population which is increasingly constipated or suffering from harder or denser stool (with low vegetable and fiber in our diets). But even old toilet designs were not optimal or efficient. In fact, humans have not improved the design of toilets for over a century. Many countries in the world do not even have access to hygienic toilets. So should we abandon space missions and instead spend research money on improving our toilets and making them available to all humans? Or should we abandon eco-efficient water-conserving toilet designs? Neither. We should just balance the perspective of our curiosity between heavenly objects and mundane earthly ones like toilets or our own bowel movements and balanced diets. Our top brains could be looking for microbiota on the surface of Mars yet not be aware of our own dysfunctional gut microbiota that make us sick and constipated. We should be more curious about our own bodies, and our homeostatic balance. Maybe a healthier diet consisting of more fiber and less meat could help us conserve water with the eco-efficient toilets.

Epilogue: A Denatured Species with an Uncertain Future

We basically face two main choices for our future as a species: Path of scalable quantity, speed and convenience (Not-Enough) or path of quality, resilience and peace (Contentment). We can continue to condition our brains like *Homo economicus* or revert back to our *Homo economicus* roots. An ancient Persian proverb states: "One word is all it takes, but it needs one open mind." By word, they mean signs, feedback and warnings. As a species, we have a lot more than one sign that we have chosen a path which has left us traumatized, imbalanced and damaged. In 2021, satisfaction with *quality* of life in the U.S. reached its lowest level since Gallup began the assessment two decades ago. Thoughtful artists and writers see a future of inequity and chaos ahead of humans, as envisioned in movies like *Elysium* (by Neill Blomkamp) and *Idiocracy* (by Mike Judge) or in books such as *Ages of Discord* (by biologist Peter Turchin).

Neurologically, as we explained, when we are in a state of imbalance, we are either perseverative and yoked to routines, or reckless and too close to the edge. These are both costly because they stress us and those around us. Chronic anxiety or stress interferes with our natural homeostatic baselines but we often self-delude by using masks, crutches and daggers to ignore natural feedback which signals trauma. We seem to be the only species that can do this delusional type of self-deception. We are masters of tools and yet have more blindspots than any other species. We mainly use our tools as crutches. Maybe we need a lot of mirrors instead.

Without feedback loops, we are like a butterfly that flies too close to a flame or a light bulb, or Icarus in Greek mythology who became too proud of his wings and flew too close to the sun.

In nature, the basis for adaptive survival is self-balancing, self-healing and symbiotic metabolic efficiency. Intelligence is agility in the face of contingency, a capacity to revise by embracing feedback, inconvenience, pain, moderation and contentment, as part of survival.

Yet for modern humans instant comfort seems boundless as long as we can afford it. We can eat anything and live anyway we choose, as long as we can afford the tools and rules (laws). It is not only our relationship to nature (including our own body) that is ruled by tools. Socially, the relation between humans, employees and employers, government and citizens, neighbors, husbands and wives, business and customers, children and parents, and nations with other nations, are now ruled by tools (surveillance, phones, pills, guns, weapons, gadgets) and rules (Laws such as labor, zoning, family, divorce, custody, international, tort, criminal, business laws). We are the only species that has commoditized and made a business out of practically every blessing in our life, including health, knowledge (education) and relationships (through social media).

We impact each other more than any other eusocial species does. We even regulate each other (through laws, rules and legislation). Yet for all that connectedness, we do not learn from the way other eusocial species handle trauma and resource management. Compare our global, country-level and local-level responses to the COVID-19 pandemic with the examples I shared about strategies of ants and bees in times of distress.

We are an impatient species, victims to the neurological *temporal discounting* and *habituation* effects of overdosing on the dopamine constantly released by our success, our conveniences and scalable productivity. Yet, in nature, practically nothing good is rushed except death and destruction. Healing, resilience and adaptive evolution take time.

The silver lining of our evolution as a concept-driven cerebral being is that we can also rewrite and unplug our brains through behavior (the root of cognitive behavior therapy) by becoming balanced, metabolically efficient and feedback oriented. Society may label those who unplug as lazy, loser or crazy. This is what John Lennon went through late in life after *scaling down* his goals and dopaminergic pursuits. He captured this sentiment in *Watching the Wheels*:

"People say I'm crazy, Doing what I'm doing, .. When I say that I'm okay, well they look at me kinda strange, 'Surely, you're not happy now, you no longer play the game', People say I'm lazy, Dreaming my life away, Well they give me all kinds of advice, Designed to

enlighten me, When I tell them that I'm doing fine watching shadows on the wall, 'Don't you miss the big time boy, you're no longer on the ball?' I'm just sitting here watching the wheels go round and round, I really love to watch them roll, No longer riding on the merry-go-round, I just had to let it go!"

If imbalance and trauma are at the root of our issues, then how do we, as a biopsychosocial species, balance and heal our minds, bodies and souls? What if like Lennon, we all decided to trade off some convenience, speed and quantity with self-imposed moderation and balance? Let's do our last thought experiment and imagine a world where one species, i.e., *Homo sapiens*, despite all its capacity to binge and splurge, chooses to practice *balance* and *moderation* in a *conscious* and *self-imposed* form without legislation or brute force.

In this world, humans would have to constantly assess their own limits, like a well-adjusted smart thermostat. The intelligent species teaches children moderation (balance) because it minimizes conflicts and waste. Everyone learns from an early age to monitor biological and social feedback and maintain their body and relationships close to homeostasis and a steady state. Transgressions become rare and so do the society's need for police officers, doctors, hospitals, courthouses and prisons.

Businesses still compete and seek profits but revise targets that lead to overworked imbalanced, unhealthy and unhappy employees. Labor laws and courts practically become obsolete. Employees, on the other hand, are balanced enough to make sure they do not slack off at work, or backbite or backstab each other. They will also maintain a healthy work-life balance so there is no need for an employer or a spouse looking over their shoulders.

Balance-minded people would balance and measure their success in life not in 1 or 2 dimensions (like high income, nice houses and cars) but by using a grade point average (GPA) system similar to those used in colleges. Because humans cannot ace multiple courses in life, they assign a weight to each course of life, as in academia, depending on its importance in our overall life and the number of hours we spend on that course. For example, we assign a weight of 8 to sleep if we sleep 8 hours a day, and assign a score to the *quality* of sleep, from A (excellent, regular deep sleep and proper sleep cycles, feeling fully refreshed in the morning) to F (insomniac). If we spend 2 hours a day for healthy metabolism (nutrition, physical activity or exercise) we will assign a weight of 2 to the course called *metabolic health* and a *quality* score of A to F depending on our overall state of homeostatic balance and health (based on biomarkers discussed in this book). We would first assign proper weight and resources (time) to all the important courses in life and then utilize resources like money and brain energy (attention) and skills to do well in each course, particularly the ones with higher weight. This grade report system ensures we pay attention to top priorities in our lives. For instance, if we say our family relations are important to us, and yet we spend twice as much time on entertainment (video games, movies, etc.), we will progress very poorly in the course of *family relations* in life. I will post a sample of this Balance Grade Report on my blogs.

In a utopian balance-oriented world, everyone seems to know their limits and how to stay close to a balanced state. Neighbors do not trespass, know their limits and keep their noise within their walls. School children are raised to balance their own affairs and not to bully or be bullied. Teachers and principals do not feel overwhelmed or overworked. Children need no reminders to do homework and chores, sleep on time, or stop texting or playing video games. Husbands, wives, roommates and partners need no nudging to remind each other of important dates and tasks or not to cheat. Negative externalities are at a minimum in such a society and economy.

In their social interaction, people are attracted to balanced, modest and honest people. People choose those political candidates not with the hottest rhetoric but with a balanced personal life and healthy bodies, friends, family and social relations. Balance is justice, and like peace, is a way of life not a slogan.

In their personal lives, people stay attuned to feedback from their body, mind and from people around. They eat when they are really (catabolically) hungry and in general watch for slippery slopes and perseverative or addictive behavior. Lives become low-drama, low-trauma and also low-cost because excess and imbalance in various forms such as obesity, disease and hoarding become rare. People buy only what they really need to remain balanced, so they choose the *quality* over quantity of purchases. Life becomes more about the quality of our experiences so *time* and *deadlines* become less of a pressure. People value exploration more than exploitation, collaboration more than competition. With less hoarding and consumption at the top, surpluses trickle down so for the first time in history, very few people live under poverty line. Crime rates drop. Everyone sleeps, lives and loves well.

Well, you just imagined what our world would look like if we were raised with personal balance as a guiding principle and adaptive evolutionary trait. It would be naturally adaptive because the large number of people living self-adjusted, cost-effective, resilient, cooperative lives would reduce the cost of trauma to the whole ecosystem. If living as a human is like walking on a treacherous tightrope (see earlier figure), why not practice self-balancing to become skillful acrobats and joyful masters of effortless balance?

Chapter References:

[1] Walter, Kimberly N, et al. "Elevated Thyroid Stimulating Hormone Is Associated with Elevated Cortisol in Healthy Young Men and Women." *Thyroid Research*, vol. 5, no. 1, 2012.
[2] Van der Spoel, Evie, et al. "Interrelationships between Pituitary Hormones as Assessed from 24-Hour Serum Concentrations in Healthy Older Subjects." *The Journal of Clinical Endocrinology & Metabolism*, vol. 105, no. 4, 2019.
[3] Tmurphy, et al. *Do the Math*, 29 Nov. 2011, https://dothemath.ucsd.edu/2011/11/mpg-of-a-human/.
[4] Scherer, Thomas, et al. "Brain Insulin Signalling in Metabolic Homeostasis and Disease." *Nature Reviews Endocrinology*, vol. 17, no. 8, 2021, pp. 468–483.
[5] Pontzer, Herman et al. "Hunter-gatherer energetics and human obesity." *PloS one* vol. 7,7 (2012): e40503. doi:10.1371/journal.pone.0040503

[6] Panossian, Alexander, and Georg Wikman. "Effects of Adaptogens on the Central Nervous System and the Molecular Mechanisms Associated with Their Stress—Protective Activity." *Pharmaceuticals*, vol. 3, no. 1, 2010, pp. 188–224.

[7] McIntosh, C, and J Chick. "Alcohol and the Nervous System." *Journal of Neurology, Neurosurgery & Psychiatry*, BMJ Publishing Group Ltd, 1 Sept. 2004, https://jnnp.bmj.com/content/75/suppl_3/iii16.

[8] Van de Wouw, Marcel, et al. "Distinct Actions of the Fermented Beverage Kefir on Host Behaviour, Immunity and Microbiome Gut-Brain Modules in the Mouse." *Microbiome*, vol. 8, no. 1, 2020.

[9] Herbert, J. "Cortisol and Depression: Three Questions for Psychiatry." *Psychological Medicine*, vol. 43, no. 3, 2012, pp. 449–469.

[10] Cabrera, Daniel, et al. "Time-Restricted Feeding Prolongs Lifespan InDrosophilain a Peripheral Clock-Dependent Manner." *BioRxiv*, 14 Sept. 2020.

[11] Koronowski, Kevin B., and Paolo Sassone-Corsi. "Communicating Clocks Shape Circadian Homeostasis." *Science*, vol. 371, no. 6530, 2021.

[12] Lindstedt, Stan L., et al. "Animal Galloping and Human Hopping: An Energetics and Biomechanics Laboratory Exercise." *Advances in Physiology Education*, vol. 37, no. 4, 2013, pp. 377–383.

[13] Akhlaghpour, Hessam "A 7 Minute Timer Has Been Discovered in Neurons." *Life Is Computation*, 8 Sept. 2021. https://www.lifeiscomputation.com/a-7-minute-timer-has-been-discovered-in-neurons/

[14] Grippo, Ryan M., et al. "Dopamine Signaling in the Suprachiasmatic Nucleus Enables Weight Gain Associated with Hedonic Feeding." *Current Biology*, vol. 30, no. 2, 2020.

[15] Horvath, Steve, et al. "An Epigenetic Clock Analysis of Race/Ethnicity, Sex, and Coronary Heart Disease." *Genome Biology*, vol. 17, no. 1, 2016.

[16] Glenn, Andrea L., and Adrian Raine. "The neurobiology of psychopathy." *Psychiatric Clinics of North America* 31.3 (2008): 463-475.

[17] Sallaberry, Chad, and Laurie Astern. "The Endocannabinoid System, Our Universal Regulator." *Journal of Young Investigators*, 1 June 2018.

[18] Gu, Xiaosi, et al. "Anterior Insular Cortex Is Necessary for Empathetic Pain Perception." *Brain*, vol. 135, no. 9, 2012, pp. 2726–2735.

[19] Love, Tiffany M. "Oxytocin, Motivation and the Role of Dopamine." *Pharmacology Biochemistry and Behavior*, vol. 119, 2014, pp. 49–60.

[20] Li, Tong, et al. "Approaches Mediating Oxytocin Regulation of the Immune System." *Frontiers in Immunology*, vol. 7, 2017.

[21] Gesquiere, Laurence R., et al. "Life at the Top: Rank and Stress in Wild Male Baboons." *Science*, vol. 333, no. 6040, 2011, pp. 357–360.

[22] Greene, J. D., and J. M. Paxton. "Patterns of Neural Activity Associated with Honest and Dishonest Moral Decisions." *Proceedings of the National Academy of Sciences*, vol. 106, no. 30, 2009, pp. 12506–12511.

[23] Useem, Jerry. "Power Causes Brain Damage." *The Atlantic*, Atlantic Media Company, 23 June 2017, https://www.theatlantic.com/magazine/archive/2017/07/power-causes-brain-damage/528711/.

[24] Fancourt, Daisy, et al. "Effects of Group Drumming Interventions on Anxiety, Depression, Social Resilience and Inflammatory Immune Response among Mental Health Service Users." *PLOS ONE*, vol. 11, no. 3, 2016.

[25] "Cody James Holland, 25, of Heuvelton." *Https://Www.wwnytv.com*, https://www.wwnytv.com/2020/08/10/cody-james-holland-ogdensburg/.

[26] Burnett, A L. "Role of nitric oxide in the physiology of erection." *Biology of reproduction* vol. 52,3 (1995): 485-9.

[27] Graf, Heiko, et al. "Serotonergic, Dopaminergic, and Noradrenergic Modulation of Erotic Stimulus Processing in the Male Human Brain." *Journal of Clinical Medicine*, vol. 8, no. 3, 2019, p. 363.

[28] Romano, Michela, et al. "Problematic Internet Users' Skin Conductance and Anxiety Increase after Exposure to the Internet." *Addictive Behaviors*, vol. 75, 2017, pp. 70–74.

[29] *Voltaire's Philosophical Dictionary*, https://history.hanover.edu/texts/voltaire/volequal.html.

[30] Darrin Qualman Darrin Qualman is a long-term thinker, et al. "Civilization as Asteroid: Humans, Livestock, and Extinctions." *Resilience*, 13 June 2018, https://www.resilience.org/stories/2018-06-13/civilization-as-asteroid-humans-livestock-and-extinctions/.

[31] Bar-On, Yinon M., et al. "The Biomass Distribution on Earth." *Proceedings of the National Academy of Sciences*, vol. 115, no. 25, 2018, pp. 6506–6511.

[32] "Measuring Collective Intelligence ." *MIT Center for Collective Intelligence*, https://cci.mit.edu/mci/.

[33] Taneja, Veena. "Sex Hormones Determine Immune Response." *Frontiers in immunology* vol. 9 1931. 27 Aug. 2018, doi:10.3389/fimmu.2018.01931

www.ingramcontent.com/pod-product-compliance
Lightning Source LLC
Chambersburg PA
CBHW020441130626
46549CB00001B/246